ALASKA GEOGRAPHIC.

Volume 18, Number 1

ALASKA'S WEATHER

The Alaska Geographic Society

To teach many to better know and more wisely use our natural resources

EDITOR
Penny Rennick

ASSOCIATE EDITOR/PRODUCTION COORDINATOR
Kathy Doogan

STAFF WRITER
L.J. Campbell

DIRECTOR OF SALES AND PROMOTION
Kaci Cronkhite

MEMBERSHIP/CIRCULATION ASSISTANT
Lori Granucci

ALASKA GEOGRAPHIC® (ISSN 0361-1353) is published quarterly by The Alaska Geographic Society, 137 East 7th Avenue, Anchorage, Alaska 99501. Second-class postage paid at Anchorage, Alaska, and additional mailing offices. Printed in U.S.A. Copyright © 1991 by The Alaska Geographic Society. All rights reserved. Registered trademark: Alaska Geographic, ISSN 0361-1353; Key title Alaska Geographic.

POSTMASTER: Send address changes to
ALASKA GEOGRAPHIC®
P.O. Box 93370
Anchorage, Alaska 99509-3370

COVER: *A double rainbow crowns the Valdez small boat harbor with Sugarloaf Mountain (3,484 feet) in the background.* (Jon Nickles)

PREVIOUS PAGE: *A storm breaks over Glacier Bay National Park and Preserve in southeastern Alaska.* (Steve McCutcheon)

OPPOSITE: *The south side of the Alaska Range corrals some of the Gulf of Alaska's moisture that in winter falls as snow. David Fox snowshoes along a creek flowing out of the High Mountain Lakes region in the foothills of the Tordrillo Mountains of the Alaska Range. River otter tracks run along the creek's left side.* (Shelley Schneider)

BOARD OF DIRECTORS
Robert A. Henning, *President*,
Judge Thomas Stewart, Phyllis Henning, Jim Brooks,
Charles Herbert, Celia Hunter, Byron Mallott,
Dr. Glen Olds, Penny Rennick

NATIONAL ADVISORS
Gilbert Grosvenor, Bradford Washburn, Dr. John Reed

THE ALASKA GEOGRAPHIC SOCIETY is a non-profit organization exploring new frontiers of knowledge across the lands of the Polar Rim, putting the geography book back in the classroom, exploring new methods of teaching and learning—sharing in the excitment of discovery in man's wonderful new world north of 51°16'.

MEMBERS OF THE SOCIETY receive the *ALASKA GEOGRAPHIC®*, a quality magazine that devotes each quarterly issue to monographic in-depth coverage of a northern geographic region or resource-oriented subject.

MEMBERSHIP DUES in The Alaska Geographic Society are $39 per year, $43 to non-U.S. addresses. ($31.20 of the $39 yearly dues is for a one-year subscription to *ALASKA GEOGRAPHIC®*.) Order from The Alaska Geographic Society, P.O. Box 93370, Anchorage, Alaska 99509-3370; phone (907) 258-2515.

MATERIALS SOUGHT: *ALASKA GEOGRAPHIC®* editors seek a wide variety of informative material on the lands north of 51°16' on geographic subjects—anything to do with resources and their uses (with heavy emphasis on quality color photography)—from all the lands of the Polar Rim and the economically related North Pacific Rim. We cannot be responsible for submissions not accompanied by sufficient postage for return by certified mail. Payments for all material are made upon publication.

CHANGE OF ADDRESS: The post office does not automatically forward *ALASKA GEOGRAPHIC®* when you move. To ensure continuous service, notify us six weeks before moving. Send your new address and zip code, and if possible, a mailing label from a copy of *ALASKA GEOGRAPHIC®*, to: *ALASKA GEOGRAPHIC®*, P.O. Box 93370, Anchorage, Alaska 99509-3370.

MAILING LISTS: We have begun making our members' names and addresses available to carefully screened publications and companies whose products and activities may be of interest to you. If you would prefer not to receive such mailings, please advise us, and include your mailing label (or your name and address if label is not available).

ABOUT THIS ISSUE: We are pleased to have one of the best-known weathermen in Alaska as the author of the main text for this issue. Mark Evangelista, meteorologist with the National Weather Service, is host of a statewide television program on weather and author of a column on weather for the *Anchorage Daily News*. We thank geophysicist Mark McDermott for the insiders' look at rainbows, and longtime Alaskan Jill Shepherd for her firsthand account of Alaska's hot weather. Editor Penny Rennick prepared the introduction for this issue, and staff writer L.J. Campbell put the finishing touches on the overall picture with her sidebars on some of the more magical elements of the state's weather environment and on how some Alaskans deal with that weather.

We are grateful to Neal Brown and Glenn E. Shaw of the Geophysical Institute, University of Alaska Fairbanks; to George Cebula, National Weather Service; and James Wise and Ron McClain, Alaska Climate Center, for invaluable technical information, photo support and review of portions of the manuscript.

COLOR SEPARATIONS BY
World of Colors USA, Inc.

PRINTED BY
Hart Press

PRICE TO NON-MEMBERS THIS ISSUE: $17.95

ISBN 0-88240-196-3

The Library of Congress has cataloged this serial publication as follows:

Alaska Geographic. v.1-
[Anchorage, Alaska Geographic Society] 1972-
v. ill. (part col.). 23 x 31 cm.
Quarterly
Official publication of The Alaska Geographic Society.
Key title: Alaska geographic, ISSN 0361-1353.

1. Alaska—Description and travel—1959-
—Periodicals. I. Alaska Geographic Society.

F901.A266   917.98'04'505   72-92087

Library of Congress   75[79112]   MARC-S

CONTENTS

INTRODUCTION

"**W**hat's the weather like up there?" How many times have Alaskans heard that question when answering a call from someone Outside. Weather is big time, worthy of an entire program on the statewide television network and a complete page in one of the Anchorage newspapers. And no wonder. Knowledge of upcoming weather can mean the difference between life and death. Extreme cold, high winds, rough seas, shifting ice can threaten Alaskans at every turn. Good judgment says check the weather before setting out, and when a television weather map has DANGEROUS COLD written in big letters across the entire western half of the state, as it did in January 1989, Alaskans stay indoors as much as possible.

In late December hoarfrost coats branches overhanging Talachulitna Creek in southcentral Alaska. (Shelley Schneider)

When they have to go outside, Alaskans have a wide variety of gear with which to stay comfortable: bunny boots, breakup boots, mukluks, the line of footgear fills an entire closet, as do the coats, pants, headgear and mittens.

For extreme cold, residents fit their vehicles with heaters and plug their cars in to warm the fluids before starting the engine. Streets of downtown Fairbanks routinely have plug-ins for keeping engines warm. Cars that sit for any time in deep cold acquire "square" tires frozen on one side, and tires that pop off their rims. In Southeast, where ice and high winds challenge pedestrians on the steep streets of many communities, ice cleats that can be attached to shoes or boots hang on a handy hook.

Heavy, wet snow, common in Southeast and in coastal areas of southcentral Alaska, means shoveling roofs before they collapse and clearing boat decks before the vessels sink under the

LEFT: *A storm throws waves onto the Pacific beach of Amchitka Island in the Aleutians.* (Steve McCutcheon)

ABOVE: *Alaskans have fun in the snow at the annual snow sculpture competition held during Anchorage's Fur Rendezvous.* (Cary Anderson)

ABOVE RIGHT: *In this photo taken at midnight, mist rises from the Hulahula River as it nears the Beaufort Sea coast in the Arctic National Wildlife Refuge.* (Chlaus Lotscher)

weight of their white blanket several feet thick. High winds require tying down the small aircraft commonly found throughout the state to prevent them from flipping. Freeze-up necessitates putting the planes on skis; breakup requires getting the planes on solid ground before they sink through the ice frequently used for runways.

"Weather permitting" is the password in Alaska, the afterthought attached to every conversation about travel or shipment. And sometimes the weather does not permit. Many Alaskans have spent hours or days huddled around the fire in a roadhouse, stretched on the

floor of a crowded airport terminal, or curled in a sleeping bag in a tent, waiting for the promised plane or boat.

As winter gives way to a brief spring, the roar of breakup thunders along Alaska's waterways. Beginning far upstream, a quickening current carries ever-growing plates of ice toward the sea. When the ice gets hung up, debris and more ice build up behind the dam, causing floods that have plagued riverfront communities since their birth. Warmer weather, especially inland away from coastal winds, brings the mosquitoes, so vicious at times that Alaskans wish for winter's cold. The warmth also means thawing, and the muck that mires vehicles anywhere off the paved roads. When the mud dries, the dust comes, trailing behind vehicles in a roostertail of grit and brown frosting. The contrast in air and ground temperature creates fog, a hazard to navigation and flying. How many times have pilots overflown the Pribilofs or St. Lawrence Island because of the fog, or skimmed the North Slope tundra searching for the haven of the Barrow runway.

But when Alaska's weather is good, it can be spectacular. Warmth, combined with long daylight, brings giant vegetables, and swimming in the local hole, even above the Arctic Circle. The surfers take to the waves off Yakutat, and the water skiers to Kotzebue Sound. Hikers, fishermen, river runners can be found just about anywhere, enjoying the 24-hour daylight and fresh air that goes with living in the land of the midnight sun.

Many Alaskans choose fall as their favorite season. The air is crisp but not cold. The mosquitoes are gone. The tundra blazes in red, while the taiga shimmers in gold. Wild animals look their finest, their thick coats ready for the coming cold. The state's human residents see to a full woodpile, a well-stocked pantry, a finely tuned engine, because they know that the state's most notorious weather is just around the corner.

What is the weather like in Alaska? Let us take a look.

UNDERSTANDING ALASKA'S WEATHER

BY MARK EVANGELISTA

Editor's note: *Mark is a meteorologist with the National Weather Service, writes a weather column for a local newspaper and hosts* Alaska Weather *on statewide television.*

Mention Tennessee to someone and ask them to tell you the first thing that comes to mind. Some may describe the people. Others may talk of Nashville, or of other cities and towns. But few would say anything about the climate.

Now try the same thing with Alaska. Suddenly the climate is a big item. Perhaps no other state,

Anchorage summer highs are usually in the 50s or 60s; a rare heat wave of 70- to 80-degree temperatures lures swimmers and sunbathers to Goose Lake on the east side of town. (Pete Martin)

with the possible exception of Hawaii, is as immediately identified with its climate as is Alaska. The association is so pronounced that the words *Alaska* and *winter* may share the same synapse in the brain.

Individual perceptions of the Alaskan climate depend on a variety of factors, not the least of which is actual time spent in the state. Many people who view Alaska as a land of eternal winter, where softball is played with snowshoes out of necessity, have only seen the state on magazine pages and television screens. Others who have visited only briefly carry misconceptions of the state's weather. Even those who have lived in Alaska for some time exhibit skewed views of its climate, some preferring a winter in the Interior to the relatively intolerable summer heat of the Lower 48.

Personal experiences and secondhand weather tales are common foundations for developing impressions of the Alaskan climate, but to truly understand the state's weather observers have to look at climatology, the weather history. Even though some of Alaska's weather records only go back to the turn of the century, a fairly short time compared to records in other states, they still contain a wealth of weather information.

This information will not be understood, however, simply by staring at pages of data because the data will merely reinforce or disprove perceptions. The numbers are pieces of a greater puzzle, the puzzle of understanding and, ultimately, of predicting the weather.

Unlocking the climatological history is certainly important to understanding Alaska's weather, but it is not necessarily a good first step. It is just as important to know what can be learned from the data, for instance how does change in humidity affect temperature, if indeed it does, before examining the numbers. The journey toward understanding Alaska's climate does not begin in the weather archives, it begins out in space.

FORCES BEHIND THE WEATHER
Planets in our solar system receive almost all of their energy from the sun. In fact, the proximity of each planet to the sun determines that planet's relative temperature: Mercury is the warmest while Pluto is the coldest. The sun's energy comes to Earth in the form of solar radiation, or sunshine. The radiation passes through the atmosphere, losing little of its energy to air molecules.

LEFT: *Frost on a cabin window is backlit by the rising sun on a cold winter morning.* (Shelley Schneider)

RIGHT: *In minus 30 degree temperatures in late December, frost crystals up to five inches in diameter form across the surface of Judd Lake in southcentral Alaska.* (Shelley Schneider)

Alaska's Climatic Regions

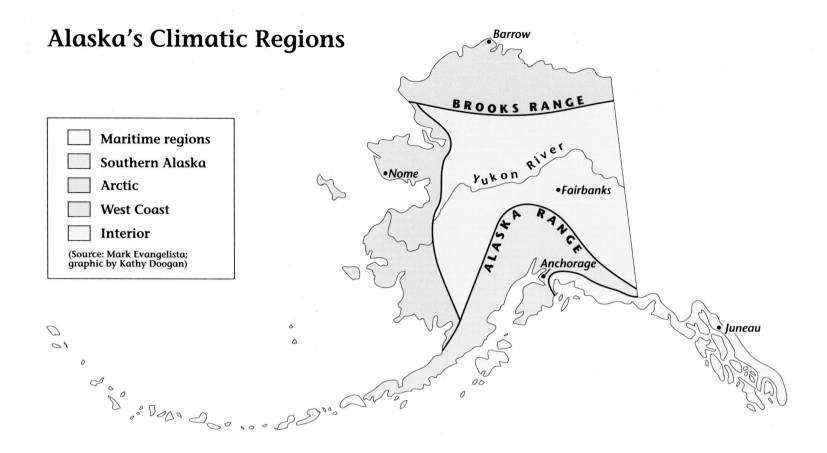

Legend:
- Maritime regions
- Southern Alaska
- Arctic
- West Coast
- Interior

(Source: Mark Evangelista; graphic by Kathy Doogan)

Barrow

BROOKS RANGE

•*Nome*

Yukon River

•*Fairbanks*

ALASKA RANGE

Anchorage

•*Juneau*

In radiation form, the sun's energy is not good at heating the atmosphere but it does a great job of heating the Earth's surface. The warm surface easily passes that warmth to the atmosphere. As far as the atmosphere is concerned, the Earth's surface is the primary source of heat. Just as proximity to the sun determines a planet's temperature, closeness to the Earth's surface usually determines the temperature of the atmosphere, the warmest air being closest to the ground.

Throughout the entire year, the Earth's equatorial region receives more solar energy than does any other part of the planet. The relationship of Earth's orbit to that of the sun exposes this region to the greatest amount of direct sunlight. Direct sunlight, that which arrives from directly overhead, passes through less atmosphere than does sunlight that approaches the Earth obliquely. In addition, the equatorial region receives a consistent measure of sunlight throughout the year. There are no long periods of little sunlight, as there are at higher latitudes.

This unequal distribution of solar radiation creates an unequal distribution of heat. Air over the equator receives much more terrestrial heat than does air over the poles. Consequently, at any given altitude equatorial air is normally warmer than air anywhere else on the globe. At some arbitrary altitude, say 10,000 feet, air at the poles is typically colder than air at 45 degrees latitude. Equatorial air at this altitude would be warmest of all.

Ice fog obscures a look down Second Avenue in downtown Fairbanks. This thick fog forms when an air inversion traps cold air close to the ground, and the lower layer becomes so cold that it can no longer hold water vapor. (Charles Kay)

Southeast: Inescapably Wet

Alaska is known as a land of extremes, and Southeast Alaska is known for extreme rain.

This narrow strip of coastal mainland and islands runs some 600 miles along the Gulf of Alaska. It is a panhandle of lowlands, lush forests and glaciated mountains, of fishing villages and logging camps, cool summers and warm winters. It is also a dumping ground for storms off the north Pacific Ocean.

Southeast has rain, and lots of it. More than 100 inches fall in much of the region each year. In the higher elevations, the moisture falls as wet snow. Southeast snow is so wet that, as a general rule, 10 inches of it equal an inch of water, compared to 15 inches or more of drier interior snow. Part of Southeast's snow turns to ice that feeds the region's many glaciers.

But it is the rain that fashions the lives of Southeasterners. Rubber rain gear goes everywhere, as do the ubiquitous rubber boots, also known as Ketchikan tennis shoes. It is the rain they joke about; four consecutive days of rain in Southeast is considered a brief shower. It is not uncommon for winter storms to bring 20 inches or more of rain a month to the southern third of the panhandle. In October, one of the region's wettest months, it rains 47 percent of the time in Juneau, 43 percent of the time in Yakutat

One of the fringe benefits of abundant rainfall is the frequent appearance of rainbows, such as this one over a main street in downtown Skagway. Located at the upper end of Lynn Canal and protected by steep mountains, Skagway is not quite as wet as some towns farther south in the region. (Harry M. Walker)

and 42 percent of the time in Annette, according to Tom Schornak, a weather observer in rain-soaked Sitka who wrote a weather column for the *Southeastern Log*, the now-defunct monthly newspaper out of Ketchikan.

The amount of precipitation does vary within Southeast, depending on the combination of bays, inlets, ridges, forests, slopes, rivers and valleys. For instance, Skagway averages only about 30 inches of rain a year, a virtual desert compared to other places in Southeast, largely because of its sheltered location at the base of mountains at the northern tip of Lynn Canal. Likewise, Angoon appears to be one of the region's drier spots. Located on the eastern side of Chatham Strait and sheltered from Pacific Ocean blasts by Baranof Island, Angoon averages only about 40 inches of rain a year. Yet, the community holds the state record for the most precipitation in 24 hours; 15.2 inches on Oct. 12, 1982. Without question, Southeast is inescapably wet. Weather stations report annual precipitation averaging from 60 inches in Haines to 155 inches in Ketchikan. This includes snow, which is melted by the weather observers to measure as water.

Ketchikan residents see the humor in their reputation for abundant rainfall. (Harry M. Walker)

One of the state's all-time rainiest spots is Little Port Walter where yearly rainfall averages 240 inches. A thousand feet above Little Port Walter on the mountainside, the precipitation is 300 inches a year.

Bob Budke and his wife, Polly, are Little Port Walter's sole inhabitants, caretakers of the National Marine Fisheries Service salmon research center here. It does not rain every day, Budke says, but when it does rain, it is not unusual for 2 to 3 inches to fall. The record rainfall for a day was 8 inches. A typical rainy spell lasts a week or longer. "It's something you get used to," he says by marine radio telephone. "When it rains you just put on extra clothing or rain gear. You don't stop work. You just put up with it."

The weather systems that bring this type of wetness to Southeast make for dicey travel. Few communities are connected by roads. Most travel, mail and freight deliveries are by the state ferry to communities along the Inside Passage, or by fishing boats and floatplanes, the only

links to more remote settlements. Flying weather is frequently and persistently poor to marginal.

Pilot Mike Sullivan has been flying single-engine floatplanes for air freight companies in Southeast for 11 years. "One thing I always tell new pilots is that Southeast Alaska is an area of mini-climates," he says. "You can be in Sitka with blue sky and fly to the other side of the island to Port Alexander and there's

Hoarfrost on a winter morning glazes these trees near Peterson Creek, 20 miles north of Juneau. (John Hyde)

I have to," Sullivan says, "But if the wind is strong and the water is too choppy to land, I have to have high ceilings."

Then there are times, sometimes two weeks in a row during winter, that weather is so bad that planes just do not fly.

That means people in places like Port Alexander, on the southern tip of Baranof Island, are virtually stranded. They do not get mail, groceries or visitors, all of which normally come by plane. They also do not go very far in a hurry. A trip up the island to Sitka, which takes about 40 minutes by air on a clear day, takes 10 hours by boat.

Just as Southeast is one of Alaska's rainiest regions, Port Alexander is one of Southeast's rainiest towns. Storms off Cape Ommaney frequently pummel the town, churning up waves that are said to look like small hotels coming ashore. Port Alexander, which is only a few miles south of Little Port Walter, averages 168 inches of rain a year.

Debra Dozier discovered what wet meant when she moved here in 1978. She

fog down to the water with zero visibility. Then you can go north to Juneau and the ceiling can raise and the wind will pick up to 30, 40 knots. That's all within 20 or 30 miles.

"It's full of little tiny weather systems. Rain here, fog there, winds another place. It's difficult to fly in Southeast."

The cardinal rule, Sullivan says, is never fly around the south end of an island, such as Baranof. "If there are any weather changes, they'll all focus on the point. You'll go from beautiful weather to turbulence, winds, fog banks, you never know." When the weather is nice, with high ceilings and good visibility, pilots take the shortest routes over the mountains. When visibility is low, they skim the water, hugging the coastline. With pontoons, a floatplane has a runway beneath it, if the waters are not too rough. "If the ceilings are low and the visibility is low, but there's no wind, it's okay. The water's calm and I can land if

left the arid desert of Phoenix, Ariz., trading in her sunsuit for rain gear. She took little comfort in Port Alexander's warm-for-Alaska temperatures that average 50 degrees in summer and rarely dip below freezing in winter. "It was difficult to come here. I wore a winter coat all the first summer," she says.

But that was 12 years ago, and Dozier's outlook has since changed. She enjoys the climate and hardly notices the rain now. "It becomes part of your life and you kind of ignore it," she says. "You expect it to rain all the time, and when it doesn't, well. . . . I wear my rain gear to work every morning, whether it's raining or not."

Debra works part time as a classroom aid in Port Alexander's school. The school is in the old section of town, across the bay from most of the homes. Debra paddles a kayak to school each day. In fact, she paddles the kayak where she needs to go around town. Her children

even gave her a rubberized purse for Christmas.

All but three of the school's 26 students live across the bay. Some of them drive skiffs to school; the others come in the school boat, an 18-foot Lund aluminum skiff. Depending on the tide, they know where to gather each morning to meet the boat. Dressed in float coats, rubber rain pants and wool hats, they climb in and sit quietly so they do not rock the boat. On days when the weather is bad, the parents confer in the mornings over

the CB radio and telephone. If the parents decide the water is too rough for the school boat to safely cross, they keep their children home and the teachers radio or telephone their assignments to them. The other three students, who live on the school side of the bay, attend as normal when this happens.

The mail plane lands three days a week, usually when the children are in school. With permission slips signed by their parents, they are allowed to go to the post office during lunch, says teacher

Southeast's moist climate is perfect for nourishing a rainforest, much of which is contained within Tongass National Forest. Bald eagles, as well as other species, rely on the old-growth giants of the forest as nesting trees. This eagle was photographed near the entrance to Windfall Harbor on Admiralty Island. (John Hyde)

Lucy Bikulcs. A stretch of bad weather without deliveries makes everyone anxious for the mail plane, particularly since most of the groceries come that way. Ten-year-old Nathan Young earns extra money by delivering the groceries in his skiff to the three teachers without boats who live on the water. Otherwise, they would have to push their groceries in a handcart from the floatplane dock down a boardwalk to their homes. Nathan recently ordered a new eight-horsepower motor for his aluminum skiff; his old three-horsepower motor and wooden skiff go to his 7-year-old brother.

Debra, her husband, Parres, and their three children own and operate a commercial fishing boat. He worked construction when they first came here to live near his sister, an artist who has since moved to Washington state. They stayed and when construction on the new school house, the water plant and the boardwalk was completed, they took up commercial fishing, occupation of the majority of the town's 100 or so residents.

Rain does not stop the fun for tourists to Southeast. Jane Eidler, owner of a Sitka tour company, provides sturdy umbrellas for her clients. (Karen Jettmar)

Port Alexander's sheltered harbor attracted fishermen who trolled Chatham Strait in the early 1900s. The settlement prospered through the 1930s, its harbor bustling with as many as 1,000 boats at a time. A decline in salmon and herring stocks, due to the damming of important spawning rivers, and the outbreak of World War II knocked the bottom out of fish buying, packing and processing at Port Alexander. By 1950, only 20 or so people lived here. In the early 1970s, federal land was opened to homesteading

and a new wave of settlers arrived, many of them attracted by the subsistence lifestyle and quiet surroundings.

Mark Kirchhoff homesteaded here in 1976. He arrived in Port Alexander from his home in Rochester, N.Y., newly graduated with a degree in biology from Syracuse University. His brother was living in Alaska at the time. They staked their parcel across the bay from town, in what is still known today as Tract B. They pitched a tent on the beach and proceeded to build a crude log cabin 150

Known as the "Voice of P.A." and 28-year resident of Port Alexander on Baranof Island, Tom Perrigo says he does not mind the rain that makes his town one of the state's wettest spots. "It never rains inside," says Tom. (Shannon Lowry, courtesy of Alaska *magazine)*

New Yorkers. They used to watch us with their binoculars from across the bay."

Kirchhoff lived here more or less full time until 1989 and was mayor for two of those years; now he splits his time between Port Alexander and Juneau. While living in Port Alexander, he would work a few months each year for the state department of fish and game and spend the remaining months working on a book about Baranof Island. Like other people in town, he collected rainwater for drinking and bathing, and dried his clothes on a line strung behind a wood stove. He did his writing on a manual typewriter by the light of a kerosene lamp, although now portable fuel-powered generators provide electricity for most of the houses. Until several years ago, the town had only one phone.

Tom and Marvel Perrigo, 28-year residents of Port Alexander, still answer the message phone for the town. Tom also is known as the "Voice of P.A." Every day, at 10 a.m. and 4 p.m., he broadcasts messages over CB radio channel 5. "Stuff like the pinochle game at Joe's or a P.T.A. meeting, things for sale, what hours the post office is going to be open," he says. They keep an open door with a pot of coffee on the stove and a library of video tapes to trade. At 70, Perrigo is retired, once from the Coast Guard, twice from the merchant marines. After extensive world travel, he brought his wife to Port Alexander, where he worked as a shipwright and on state ferries for a while. Now, they enjoy the peace and quiet and the colors of the sky changing over Cape Ommaney as the weather blows in.

"Right now, it's 40 degrees and sunshiny," he said one January day. "It could change in 15 minutes. But heck, you can dress for the rain. You just get so you wear boots all the time.

"As far as fishing is concerned, weather is a main thing. When you're on a boat, it's critical. If the weather isn't good, you can't make a living or you take a beating for a few bucks.

"But as far as we're concerned, it doesn't make any difference. It never rains inside."

yards or so farther inland. "The first three days were sunny. We couldn't believe our good fortune to homestead in this beautiful, beautiful spot," Mark recalls. "Then it rained for the next month. One night the tide came up past the tent." The deluge ruined his brother's camera, cached in a waterproof Army footlocker that was not so waterproof.

"As it turned out," Kirchhoff says, "the people had a lot of fun poking fun at us

LOW PRESSURE SYSTEMS

1. Since pressure is higher outside the low, wind flows into a low from all sides.

2. With nowhere else to go, air is forced upward.

3. At higher elevations, air may actually flow outward depending on pressure at those elevations. If the rising column of air has no avenue of flow outward, it continues rising, strengthening the low.

Rising air cools, and at some temperature, will be too cold to hold all its moisture. Some of the moisture will be released as clouds and precipitation.

(*Source:* Mark Evangelista; graphic by Kathy Doogan)

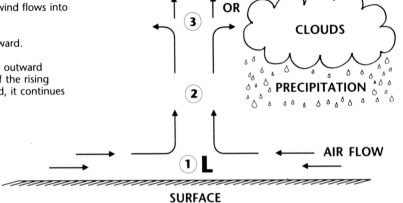

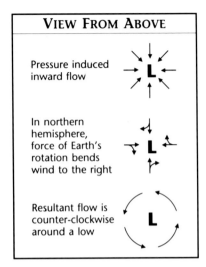

An understanding of density indicates that cold polar air is most dense and warm equatorial air least dense. Therefore, polar air, being the heaviest, would sink beneath lighter air. Conversely, lighter air would fill the upper-level space vacated by the sinking air. This rising and sinking result in a general movement of air from poles to equator at the Earth's surface, and an upper-level movement of air from equator to poles, the atmosphere's general circulation pattern.

In this pattern, air over the poles sinks to the surface, and eventually winds up at the equator where it is warmed by terrestrial heat. Once warmed, the air becomes lighter and rises. The light air migrates to the poles, replacing sinking air. As the air moves away from the equator, it migrates to regions of less terrestrial heat.

Consequently, the air cools as it nears the poles. Eventually it again becomes cold, dense air, sinks to the surface and repeats the process.

This process would seem to solve an atmospheric paradox: Air closest to the Earth's surface is warmest, while air farthest away is coldest. Gravity, however, demands that the most dense air, which would be the coldest, be closest to the surface and that the least dense, the warmest, be farthest away. General air circulation, with its continuous warming and cooling, is forever trying to catch up to the gravity-heat trade-off.

This circulation model outlines what is necessary for efficient heat exchange, but it is not even close to what actually goes on in the atmosphere. In reality, air movement is confined to much smaller areas. Small-scale processes, such as low pressure systems, operate within larger frameworks to move air masses. The idealized general circulation would not be efficient enough to effect the tremendous daily energy transfer needed to balance gravity and heat distribution. The atmosphere must use other means to transfer great amounts of energy from the equator to the poles. That is where water comes in.

Air can carry many things, but by a wide margin the most common passenger is water. By moving water from one place to another, the atmosphere transfers large amounts of energy, much more energy than the general circulation of air alone could move. Water is transported in all three physical states: solid, liquid and vapor. Vapor is the most common form, and is

normally unseen in the air; clouds and fog are made of ice or liquid droplets. Energy is transferred when water changes from one physical state to another.

Warm air can carry more water vapor than cold air. Warm air easily absorbs water from either the liquid or solid state. Large amounts of energy must be expended, however, to convert water into the vapor state. This energy remains within the vapor's molecular framework; energy is this form is known as latent heat.

If the warm, moist air is transported to a colder location, the air itself cools. As the air cools, its water capacity is reduced, eventually to the point that some vapor will be released. When this happens, the water reverts to a liquid or solid state depending on the air temperature.

HIGH PRESSURE SYSTEMS

1. Air within high-pressure column is usually more dense than surrounding air. This could be from any number of atmospheric conditions. By far the most common is the air within the column being colder than surrounding air.

2. The downward moving air slowly heats as it approaches a warm earth. This gives rise to warm summer temperatures associated with highs. Downward rushing heavy air makes surface pressure so high.

3. With nowhere else to go, downward moving air spreads out at the surface.

Since air moving within a high is continuously warming, there is no opportunity for moisture to be released and form clouds. The clear skies common to high pressure also result in cold winter temperatures, temperature inversions, fog, smog, etc.

(*Source:* Mark Evangelista; graphic by Kathy Doogan)

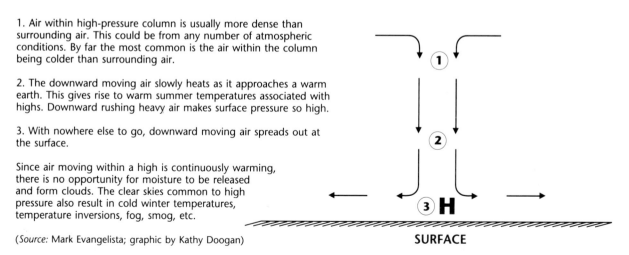

SURFACE

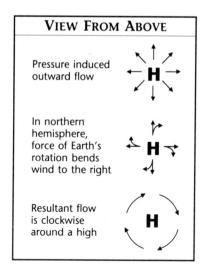

VIEW FROM ABOVE

Pressure induced outward flow

In northern hemisphere, force of Earth's rotation bends wind to the right

Resultant flow is clockwise around a high

As water changes from the vapor state, latent heat is released. The same amount of energy used to vaporize a measure of liquid water will be released when that vapor is again liquefied. The end result is a deposit of heat energy in the colder location.

The manner in which the atmosphere moves energy from the equator to the poles, by moving warm air and transforming water, results in what is known as weather. Just about every weather phenomenon has its roots in the atmosphere's basic need to transfer energy from those places that have it to those places that do not.

Left: *Santa Claus would have no trouble finding his way to this home in Anchorage.* (Pete Martin)

Right: *Gulls and bald eagles seem unperturbed by falling snow while searching for morsels of spawning chum salmon in the Chilkat River near Haines. Although one of the drier areas of Southeast because it is protected by mountains, Haines and the rest of the region receive abundant snow and rain from moisture-laden Gulf of Alaska winds.* (John Hyde)

ALASKA'S PLACE IN THE AIR CIRCULATION MODEL

Now that the forces behind the weather have been explained, how does Alaska fit into the picture. By applying general atmospheric principles, Alaskans can get their first impression of what the state's climate should be like.

Alaska's most obvious weather-influencing characteristic is its northern position on the globe. Being so close to the North Pole leaves Alaska on the debit side of the planet's energy distribution. On its own, Alaska does not receive enough solar radiation to maintain its temperature. It must depend on energy transferred from the equator. Therefore, based on its proximity to the poles, Alaskans can initially assume that their state has a cold climate, with the more northern regions being even colder.

Taking this knowledge one step further, physics says that a warm Earth will transfer some heat to the air, raising the air's temperature. Physics also says that if water is present, the newly warmed air will absorb moisture, consuming large amounts of heat energy that could have been used to further heat the air. If no water is present, no energy will be devoted to water vaporization; it will all be used to heat the air. Therefore, for the same amount of available heat energy, moist air will not warm as much as dry air will.

ALASKA'S UNIQUE WEATHER OBSERVER PROGRAM

ALASKANS SCATTERED THROUGHOUT THE state share a daily job. They watch the sky for the federal government. They look for fog, snow, rain or clear sky. They take the temperature, dew point, barometric pressure and wind speed. They fill a balloon with helium and clock its rise with a stopwatch to estimate the height of clouds overhead. Eight to 10 times a day, every day of the year, they do this.

Their information is used in advising pilots and mariners and in forecasting local weather. It is part of a federally funded program unique to Alaska in which far-flung folks are trained and paid to take weather observations for the

Instruments to measure temperature are housed inside a radiation shield (left) and gauges to measure precipitation are contained in the cone-shaped housing in this yard at Eagle on the Yukon River. The radiation shield enables the instruments to measure ambient temperature rather than super-heated direct sunlight which would give a misleading reading. (Steve McCutcheon)

National Weather Service (NWS) and the Federal Aviation Administration (FAA). They include people like Levi Seppilu, who lives in Savoonga on St. Lawrence Island in the Bering Sea; Kay Shepherd in Whittier on Prince William Sound; and James Ballweber, a minister in the interior town of Nenana. In Petersburg in southeastern Alaska, the residents of the senior citizens' home are the government's official weather watchers. All told, the program has weather observers in more than 50 locations throughout Alaska.

The paid weather observer program supplements weather data from 35 federally operated weather stations and another 75 volunteer weather observers who take temperature and precipitation readings once a day. It is all part of an effort to help government meteorologists get a more precise handle on the weather in this geographically diverse state. Many Alaskans depend on accurate weather information. Alaska has more pilots per capita than any other state, and the most practical way to reach many rural areas is by airplane. Fishing is the state's second largest industry, and boating is an important means of transportation between towns and settlements along Alaska's coasts and rivers.

The federal government spends more than $700,000 a year in payments to weather observers alone. The observers get $4.30 for daytime observations and $6.28 for those made at night. Most telephone the FAA with their observations; some in the more remote sites send their weather observations by radio transmission, a method known as "meteor burst." They type the weather data into equipment that converts the information into radio signals, which bounce off ionized meteor trails in the atmosphere to a receiver in Anchorage. The information then goes over telephone lines to FAA and NWS computers.

This box with louvered sides, known as a thermoscreen, housed weather instruments a few decades ago before the latest scientific advancements made the instruments more compact. (Steve McCutcheon)

A wind bird tops this wind tower at Icy Cape on the Chukchi Sea coast. Information gathered at this remote NOAA (National Oceanic and Atmospheric Administration) weather station is transmitted by satellite to Anchorage. (David Roseneau)

Each weather observer receives about 40 hours of training by the NWS. Usually, a weather service employee travels to the observer's home to do the training and to set up the equipment, which the government also provides.

Eventually, the paid observer program will be phased out. The government wants to go to a completely automated weather observation network by 1995. Plans call for more than 100 automated stations to be put into service around the state, many in locations where paid observers now take readings. The automated stations, shaped something like metal igloos, use a system of sensors including a visibility monitor; a laser ceilometer that bounces a beam off clouds to measure their height; a hydrothermometer to take temperature and dew point readings; an anemometer to measure wind speed and direction and an electronic barometer to measure air pressure. The automated stations will provide readings 24 hours a day. They also are powered by electricity, which means that many of the people who now do the observations in remote locations will probably end up generating the power required to keep the stations on line.

On the other hand, warm, dry air will lose its heat to a colder environment, and it will continue to lose heat until the air and environment are the same temperature. Moist air will likewise lose heat to a cold environment, but as the moist air cools, it will also release some of its moisture, and in the process, some of its latent heat will escape. Most of the latent heat will re-warm the air, resulting in an overall slower cooling than dry air. This process combined with the slower heating process described in the previous paragraph indicates that moist air cools and warms at a much slower rate than dry air.

To put the relationship another way, moist air requires a greater energy change to exhibit the same temperature changes as does dry air. The easiest way to make this relationship work in

Travelers in Alaska's Bush come prepared for all types of weather. These campers along the Tlikakila River in Lake Clark National Park wear lightweight ponchos to fend off summer showers. (Chlaus Lotscher)

understanding Alaska's climate is to state that the temperature of dry air fluctuates to a greater degree than does that of moist air. Moisture in the air tends to moderate temperature fluctuations. This is fundamental to understanding climate.

To put this principle to work, weather watchers could assume that regions of Alaska that are close to water, such as islands, peninsulas or coastlines, have a smaller temperature fluctuation than do the more land-locked parts of the state.

Before leaving this principle, weather observers must also examine a situation common to northern latitudes. In winter, most of the water sources in Alaska are frozen solid. While ice can be directly converted into vapor, the process requires tremendous energy. That energy is less available in winter because of decreased absorption of solar radiation by the Earth's surface, a result of snow reflecting much of this radiation. This situation produces an apparent loss of moisture sources once winter has left them frozen. For purposes of climate analysis, frozen bodies of water can be treated as land. Some parts of Alaska, particularly the arctic coast and northern Bering Sea coast, have two distinct climatic identities, depending on the season.

The last fundamental element of climate is terrain. Perhaps no where else on this continent is there such a concentration of distinct geographical features. Within a few miles, 10,000-foot-high mountain ranges give way to valleys that cradle frigid air for weeks at a time.

Hoarfrost and a low-angled sun create a winter vista of Alaska's largest city, Anchorage, where the average temperature in January, the coldest month, is 13 degrees. (Harry M. Walker)

High peaks rise out of warm-water bays in one corner of the state; flat, soggy marshes stretch beyond the horizon in another.

Up to this point, this survey of the climate has been intuitive. Now it becomes detailed. Nothing happens in the atmosphere without affecting something else. Alaska's rigorous terrain disrupts every weather event it touches, creating chain reactions felt across the state. Predicting how the Alaskan terrain will affect weather systems is the most difficult part of forecasting in this state. Geography has such an effect on Alaska's climate that it is practically impossible to describe its influence in general terms.

Outpost Umiat

For the past 16 years, O.J. and Elly Smith have lived, worked and operated a government weather station at Umiat. This tiny arctic settlement on the Colville River, some 75 miles inland from the Beaufort Sea, has the dubious distinction of being one of the nation's coldest spots. It frequently shows up in national weather charts with the lowest temperature for the day.

Umiat's winters bring two months of darkness and temperatures that ride between minus 30 and minus 50 for weeks at a stretch. That type of cold spurs all sorts of headaches. Mercury in the thermometer splits, which fouls accurate readings. Airplanes do not fly, which stops the flow of visitors, mail and supplies. Diesel fuel congeals, a particularly bothersome detail since the Smiths depend on diesel-powered generators for electricity and heat.

"We're pretty much busy with survival type things. We're totally self-sufficient," says Elly, a 62-year-old grandmother.

Ed Crain (left) and Pat Valkenburg, Alaska Department of Fish and Game employees, relax outside the Umiat Hilton, operated by the Smith family. (Sverre Pedersen)

"Our primary interest is keeping the generators running. If they die, that's it."

The Smiths moved to Umiat in 1975, when oil drilling was going full tilt on the North Slope. The first test well in Naval Petroleum Reserve #4 had been drilled in Umiat in 1945, when oil exploration was getting underway. In the ensuing years, Umiat grew into an important inland staging area and a major air field between Barrow and Fairbanks. The Smiths and a partner, Bill Bubbel, set up Umiat Enterprises, a multifaceted business to generate power, provide air charters and take weather observations. The Smiths built a year-round camp with ATCO trailers and began generating power for the airstrip's lights and rotating beacon, the Alascom satellite earth station and navigational radios. Their air taxi fleet included a number of prop planes and turbo props. The Smiths soon bought Bubbel's share of the company. During the winter, six to nine months a year, the business had more than 50 employees who were certified weather observers and provided around-the-clock observations and aviation advisories at Umiat and a dozen or so drilling sites in the region.

Today, the Smiths, their two grown sons and several part-time employees work the business. They still supply power to the state-owned airstrip and the satellite station, and help monitor air traffic in the region. They still run an air taxi service, though with a smaller fleet. They do guiding and outfitting during hunting seasons. And they still take Umiat's official weather observations for the National Weather Service and the Federal Aviation Administration.

After this many years in Umiat, the Smiths take the rhythms of winter in stride. The cold requires them to be constantly on the alert to potential problems. By late January, the sun starts peeking over the horizon in brief appearances that rapidly lengthen into summer's fling with continuous daylight. Ice thick enough to support airplanes still covers the lakes as late as June. But flowers are blooming by July. August brings a couple of frosts, then a brief Indian summer. "It's beautiful," says Elly. "I can't believe the variety of colors — magentas, golds, several shades of greens and yellows. Fall is absolutely gorgeous." Temperatures drop into the 20s at night, rising to 40 and 50 degrees during the day. The crisp nights bring a welcome end to mosquitoes. The return of dark nights by September brings increasingly colder temperatures, and by November the mercury is dropping well below zero again.

Umiat is designated as an emergency airstrip, and the Smiths' weather observations are important to pilots, particularly those in small aircraft. Conditions can change abruptly, which makes their job even more critical. For instance, one bright, sunny day in May the airstrip was buzzing with planes. Suddenly, winds started blowing from several directions and blitzed the field with blinding snow. The Smiths immediately radioed a warning to incoming pilots.

Even in July in Umiat, winter is never far away. One July day in the early 1980s, the Smiths were hosting a barbecue for a large group of visitors. They had flown in a pig for the occasion, and it was roasting on the spit outside. It was a scorcher of a day, about 90 degrees. Suddenly, a cold wind whipped through, and snow started falling. The party moved indoors, and the pig went into the kitchen ovens to cook. "By the time we got the pig cut up and into the big ovens, the wind had settled and it was in the 70s," Elly says. "It was the most extreme weather I can recall." ◭

The best way to examine the climate is to divide the state into sections that share similar climate-influencing features. Separated in this manner, Alaska breaks into five main parts. The following discussion will look at each part, develop ideas about that particular climate and then examine the average weather data for a location within that section.

CLIMATE IN ALASKA'S MARITIME REGIONS
The first region includes those parts of Alaska that are affected by water throughout the year. The Aleutian Islands, Kodiak, Prince William Sound and southeastern Alaska, an area known as the panhandle, fall into this region because of the relationship between their moderate weather and the nearby ocean. In these areas, even in

LEFT: *Thick banks of clouds blanket this stretch of the Inside Passage as a tug and barge ply the waters between Sitka and Juneau.* (Shelley Schneider)

BELOW: *Sonja Lotscher, 4, peers through an archway of icicles hanging from her family's log home on the bluffs above Homer. The community, on Kachemak Bay, records average low temperatures of about 20 degrees, although record lows have reached as much as 40 degrees colder.* (Chlaus Lotscher)

LOWER RIGHT: *Ground fog obscures the coast of St. Lawrence Island in this view from the Punuk Islands in the Bering Sea. Fog can disrupt travel in the region, forcing pilots to overfly island airstrips.* (David Roseneau)

winter, the water rarely freezes over. This group includes almost every Alaskan island, excluding only those that become icebound in winter.

Based on an understanding of how geography and physics affect climate, some immediate assumptions about this region's climate are possible. The panhandle is the southernmost region of Alaska, so this will be the area of most consistent year-round daylight. Consistent daylight gives the equatorial regions its surplus of heat, so the same relative consistency will grant this region steady, warm temperatures.

Water moderates temperature. This fact reinforces the idea of a steady, warm, annual temperature. This pattern certainly applies to the average monthly temperatures at Annette, on Annette Island near Ketchikan in southeastern Alaska. Annette's average temperatures remain above 30 degrees throughout the year, rising and falling with the amount of daylight. The greatest amount of daylight occurs in June, but the highest average monthly temperature is recorded in July. A similar discrepancy takes place in January, when the lowest average monthly temperature is recorded one month after December's minimum amount of daylight.

This effect, common to climates worldwide, represents the time delay necessary for the atmosphere to catch up to the gain or loss in light.

BELOW: *Snow geese follow just behind the moderating weather as they work their way north from wintering grounds along the Pacific coast to breeding grounds on the Soviet Union's Wrangel Island. This flock was photographed near the mouth of the Kenai River. (Cary Anderson)*

RIGHT: *After freeze-up, the Bering Sea becomes like an inland plain, losing its moderating effect on western Alaska's weather. Consequently the region's communities, such as Tununak on Nelson Island, endure severe cold and high winds for much of the season. (Harry M. Walker)*

Rainbows: Clues to the Colorful Sky Arcs

By Mark McDermott

Editor's note: *Mark, a geophysicist with ARCO Alaska Inc., has from time to time contributed photos and articles to* ALASKA GEOGRAPHIC®.

THE HISTORY OF THE THEORY OF THE rainbow is a story of a few brilliant minds making insightful advances, each separated by centuries of time. Aristotle seems to be the first person to have correctly deduced some of the nature of the rainbow. He surmised that it was a circular cone of some peculiar sunlight reflecting at a fixed angle, which he called "rainbow rays." Aristotle understood that the rainbow was not at a fixed point in the sky, but was visible in a certain direction, depending on the observer's position.

Nearly 17 centuries later the next significant advance in rainbow theory occurred when, in 1304, Theodoric of Freiberg, a German monk, discovered that each raindrop could form a rainbow. This same observation was rediscovered independently three centuries later by Rene Descartes. Both men used spherical flasks filled with water to trace the paths of light rays.

Not until 1666, however, did science have an adequate explanation for the most striking of the rainbow's attributes, its colors. Isaac Newton, in his famous

A double rainbow crowns the sky above the Hulahula River in the Arctic National Wildlife Refuge. The brighter primary rainbow and the fainter secondary one are formed in a similar way, but the rays of light are reflected twice in a secondary rainbow rather than just once. (Chlaus Lotscher)

A rainbow rises from low forested hills on the north side of the Alaska Range near Stampede. (Mark McDermott)

prism experiments, observed the effect known as dispersion. He found that sunlight is made up of different colors, each of which refracts, or bends, at a slightly different angle. Should pure sunlight undergo refraction, it would be separated into its component colors: the rainbow.

Only the basic principles of reflection and refraction are necessary to understand the majority of rainbow phenomena. The law of reflection states that the angle of incidence of a light ray striking a surface is equal to the angle of reflection. The law of refraction is a bit more complicated, but still easily understood.

The speed of light depends on the nature of the substance through which the light travels. Light moves considerably slower through water and glass than through air; it moves fastest in a vacuum. Because light slows down as it passes from air to water, it bends, or refracts.

The next point to consider is what occurs when rays of near parallel light strike a spherical raindrop. For simplicity, this discussion will only consider rays entering the drop along a plane through the center and parallel to the incoming rays. Significant to the formation of a rainbow is how far from the central axis (a line passing through the sun and the center of the drop) the light ray strikes the drop.

As a ray of light hits the center of the drop, traveling along the central axis, some light is reflected off the front surface and some light penetrates to reach the rear surface of the drop. The light penetrating the first surface of the drop is refracted or bent.

The light ray reaching the rear of the drop is both reflected and transmitted, but it is the reflected light that pertains to this discussion. This ray will continue back across the drop, through the center, and exit the front surface the way it came in, with no apparent refraction. At each surface the ray splits into a reflected and refracted ray, each less bright than the original. In the case of this particular ray, it encountered three surfaces before exiting the drop and was split three times. This type of ray, one that enters the drop, reflects off the back surface once and then exits the drop, is called a primary ray.

Tracing rays of sunlight entering the drop farther and farther from the central axis towards the edge of the drop, observers find that the rays bend more and more. This trend continues up to the primary rays entering the drop nearly at its edge. At this point, maximum deflection of the incoming primary rays will occur, and the rays will exit the drop at about 42 degrees from the central axis. Primary rays that enter the drop close to the edge will be bent back as they exit into the region of less than 42 degrees from the central axis. Thus incoming rays

A river runner pulls his boat along the Ukak River in Katmai National Park during a rainstorm. From the traveler's perspective, the rainbow in the distance is 42 degrees from the anti-solar point, or the point exactly 180 degrees from the sun.
(George Wuerthner)

from a wide arc are concentrated into the area around 42 degrees from the central axis. The droplet acts as a crude lens to focus the light in a specific direction.

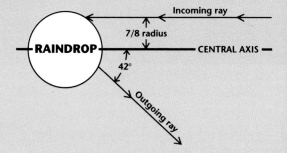

During a sky full of rain, observers looking exactly 42 degrees from the anti-solar point (the point exactly 180 degrees from the sun) will see a rainbow. If dispersion did not occur this portion of the sky would simply be bright white, but since each color is refracted at a slightly different angle, this bright area is dispersed into the colors of the rainbow. Since the water drops are spherical, they focus the primary rays in a cone 42 degrees from the axis with the sun. The only reason observers on the ground see an arc instead of a complete circle is that drops are not suspended and illuminated everywhere 42 degrees from the anti-solar point. Observers in an airplane are not under this handicap and can see the full circle.

Thus the primary rainbow results from the focusing of rays of sunlight undergoing just one internal reflection in a raindrop into a region 42 degrees from the central axis of the drop. Dispersion causes this focused light to be split into the colors of the spectrum. Since red light is diffracted less than blue light, red is able to reach the maximum angle of the primary rays making up the rainbow. As observers look at a rainbow, red light will curve around the outside of the bow since this is a greater angle (indicating less bending) away from the anti-solar point than is the inside.

Fainter secondary rainbows are formed in a similar manner; however, the light rays undergo two internal reflections, not just one. Just as the water drop focuses primary rays at 42 degrees, it concentrates secondary rays at 50 degrees with the same resulting dispersion that gives rise to the secondary rainbow's

colors. Because secondary rays are refracted such that they sweep around exactly opposite from primary rays, the colors will be reversed. Carefully tracing the path of a secondary rainbow, observers see that red attains a slightly smaller angle than blue light because it is refracted less. Thus red will appear on the inside of a secondary bow.

Since primary rays cannot be refracted at an angle greater than 42 degrees and secondary rays cannot be refracted into the region less than 50 degrees, an eight-degree portion of the sky is left where no primary or secondary rays will enter. This area appears noticeably darker than the surrounding sky and is between the two rainbows. Even when the secondary rainbow is not visible, the primary bow appears to have a dark side (the outside). This region is known as Alexander's Dark Band, after Alexander of Aphrodisias who first noted it about A.D. 200.

Why isn't a tertiary bow visible, one resulting from three internal reflections? Multiple internal reflections do occur, but since every internal reflection in a drop seriously degrades the intensity of that particular light ray, after two reflections the light is so dim that

observers do not see a tertiary bow.

When the sun is greater than 42 degrees above the horizon, no rainbow is visible from the ground. As the sun approaches the horizon a more complete rainbow is visible, culminating in an entire half circle as the rainbow is just on the horizon. Because rainbows are favored by low sun angles, visibility of rainbows is greatest at high latitudes and either early or late in the day.

So far all the major aspects of the rainbow have been nicely explained by geometrical optics. For centuries, however, one aspect of rainbow phenomena could not be understood because geometrical optics could not explain it. This is the presence of faint alternating pink and green bands on the inside of the primary bow, known as supernumerary arcs. These occur because light not only acts as rays, it also acts as waves, and hence interferes with itself, just as waves in water do. This nature of light was not appreciated before the 19th century. ◈

A rainbow curves out from the massive front of Hubbard Glacier in Disenchantment Bay. No rainbow is visible from the ground when the sun is more than 42 degrees above the horizon. (Karen Jettmar)

LEFT: *David Fox catches a ling cod, or burbot, through a hole in three feet of ice covering the surface of Judd Lake in the Susitna Basin. Ice fishing, crosscountry and downhill skiing and snowshoeing attract Alaskans looking for outdoor activities in winter.* (Shelley Schneider)

RIGHT: *A winter storm ends on the Kenai Peninsula. The region's topography creates a mixture of climates for the peninsula. Moisture patterns in the mountainous terrain of the east and south behave similarly to those in southeastern Alaska; Kenai and the northwestern quarter of the peninsula have a drier, more continental climate because flat terrain here does not break up the moisture flow.* (Michael Speaks)

So far, this discussion has not mentioned precipitation. That is simply because precipitation depends on geography. Precipitation can vary greatly from one side of a mountain to another; it is easier to calculate one temperature range for a region than to assign one average precipitation value.

In the panhandle, almost every weather system approaches from the ocean. Most of the time, powerful low pressure systems carry weather fronts into the panhandle, hitting the region with rain and snow, and strong wind. During these storms, precipitation is distributed fairly evenly throughout the area. But when major systems are absent, showers fall mostly on mountainsides that face the water. As showers pass to the mountain's other side, drier air slides down the mountain, delivering a warm "downslope" wind to the mountain's lee side.

A different type of wind blows in the northern panhandle. In this region, large glaciers cool the air. Driven by gravity, the cold,

dense air flows down the glacier's slope, delivering a frigid blast to the area below. These glacier winds, known as taku winds, can occur whenever large storms are absent.

Prince William Sound records some of the nation's heaviest precipitation. Steady onshore wind flow, driven by the many lows that move through the Gulf of Alaska, brings warm, moist air to cold mountainsides rimming the sound. Snow is routinely measured in terms of feet by the middle of a typical Valdez winter. With the record snowfall of the 1989-90 winter, Valdez residents used a snow berm as the screen for an outdoor theater. Sound was broadcast by a local radio station.

Alaska's other island groups, at least those that are not icebound, have weather similar to that of Annette Island. With few exceptions, these islands lack the tall mountains that contribute to high yearly rainfall.

HEAT: ALASKAN STYLE

BY JILL SHEPHERD

Editor's note: *"There are many different kinds of hot," responded* Alaska *magazine senior editor Jill Shepherd, when she was asked about hot weather in the Interior. An* Alaska *resident for 40 years, Jill knows heat, Alaskan style.*

FAIRBANKS IS NOT ONE OF THOSE ALASKA places where the sun does not shine. Even in deepest winter the unrelenting dark is brightened for a few hours a day. But what I found remarkable when I first moved to this interior city more than

30 years ago was that the winter sun has no Btus.

You can sit in the sun in December and never get warm. The rays are weak, almost mewling, and seem to have barely enough strength to reach Earth. But quietly, almost unnoticed, the sun begins to strut its stuff and by the middle of March is doing more than just lighting up the sky.

In March you can sit behind glass in your house or car and feel the first warmth from the sun, feel it warming the house — solar assistance with the heating bill. South-facing windows that have been frosted for months begin to thaw. I used to soak up the flood with towels I rolled up and placed on the window sills.

The first hot weather, maybe 80 degrees, can strike without warning as early as April, jolting the Interior into a quick thaw. Suddenly it is shirt-sleeve weather. We used to say Fairbanks is the only place where you can break through

An afternoon thunderstorm, characteristic of summer in the Interior, threatens showers over Fairbanks. The Interior, because of its distance from the ocean, routinely records the state's highest temperatures. Summer readings in Fairbanks average in the 60s, but can reach as high as the upper 90s. (Pete Martin)

the ice and get mud on your boots, while the wind blows dust in your eyes.

Sometimes the spring weather disappears again. I remember one spring that started off with bright, hot days only to turn cold again. The puddles refroze, and on May 21 a big blizzard laid down about a foot of snow. Earlier we had watched as a flock of 31 lesser sandhill cranes and some assorted ducks flew in from the south. They settled in the potato fields surrounding our house to eat the spuds missed by last year's potato pickers. The birds munched and rested for about a week and then headed northwest in the direction of the Minto Flats marshland. How disappointed we were to see the birds come back during the blizzard. Their return from the nesting grounds seemed to signify the return of winter.

Hot days in summer, with temperatures that can soar above 100 degrees, chased the mosquitoes and my kids into cool grass and shady woods. Our sled dogs, panting from the heat, would bury themselves up to their ears in deep holes they dug in the cold ground. None of our cars ever had air conditioning, but I do not known why. Perhaps it was just traditional to cook in your car, or the

Athabascan Indian children swim in the Yukon River at Fort Yukon on the Arctic Circle. The community gained the record as the state's hot spot when 100 degrees was officially logged in here on June 27, 1915. (Steve McCutcheon)

local car dealers figured that trying to sell an air conditioned car to an Alaskan was a slow way to get rich.

Passive solar energy has great applications in Fairbanks, at least for five months out of the year. I once designed a cabin for a solar energy class project. Based on my calculations, my 24- by 40-foot house was small enough to be 100 percent solar for 150 days, from May through September.

Even with plenty of solar radiation to heat a small house and provide domestic hot water for one or two people, the soil in Fairbanks never warms up. With an ambient temperature of 85 degrees, the soil temperature just a few inches below the surface can be as cold as 40 degrees. This explains why, on a hot day, it gets cold as soon as the sun goes behind a cloud. I learned to keep a sweater handy when I was outside in the garden, especially if I was dressed in shorts and a halter top. When clouds block the sun's warmth, a cool wind whips up suddenly,

only to subside as soon as the sun shines again.

I remember the towering cumulus clouds that formed in the sky on hot summer days. The cloud buildup usually was followed by a cold, violent rainstorm. More than once I was forced to pull my car off the road because my windshield wipers would not work in the driving rain. My daughter became hypothermic on an 80-degree days when she was caught in a storm while bicycling. This is an expected danger in the mountains, but Fairbanks is only 434 feet above sea level.

SOUTHERN ALASKA
BEYOND THE MARITIME BELT

The rest of southern Alaska can be considered another area of like climate. Except for the Alaska Peninsula, this region lies well inland although still within reach of moisture carried by frequent storms in the Gulf of Alaska. In most of this region, water only slightly affects temperature ranges, but provides enough thermal stability to prevent large-scale temperature changes.

Precipitation, again determined mostly by local terrain, is much lower than in the maritime belt. Because of the rough terrain and little intense solar energy, sizable convective storms and associated heavy rainfalls seldom occur. As a result, yearly rainfall in this part of Alaska is much lower than in land-locked states such as Oklahoma. This disparity gives rise to the phrase "northern desert."

Most bodies of water, including rivers and streams, in this region freeze solid during winter. While this freezing does cut down the amount of available moisture, there is more than enough water delivered by storms from the gulf to compensate for the frozen lakes and rivers.

And because this region lies farther north than the panhandle, the seasonal change in daylight is more drastic. Concurrently, the annual temperature fluctuation should reflect the daylight cycle. Based on an understanding of their effect on climate, as well as the effect of gulf storms, weather observers could anticipate seasonal temperature ranges to be a little larger than those in the panhandle.

While winter can definitely be cold, the average monthly temperatures during this season are well above zero, another benefit of being within reach of Gulf of Alaska low pressure systems. Warm downslope winds, similar to those in the panhandle, can also temper winter's chill.

Cold Can Be Dangerous

It is so cold when Gordon Ito pulls sheefish from the frigid waters of Hotham Inlet near Kotzebue that the fish freeze into 25- to 40-pound blocks of ice as soon as they hit the air.

Ito knows cold. Winters are frigid and blustery along the western Alaska coast, with subzero temperatures and winds in

Gordon Ito carries one of the 24 sheefish he caught this day in February 1991 off Pipe Spit at the mouth of the Noatak River. Gordon picks the fish with his bare hands to avoid damaging the net, and cleans them before they freeze. (Jim Huenergardt)

excess of 45 mph. Ito, 27, works outside every day. He fishes for salmon in the summer, sheefish in the dead of winter. In the spring, he hunts for belugas off Elephant Point to the south and Kivalina to the north. He hunts for bearded seals and polar bear among the ice floes in Kotzebue Sound, crawling over ice in camouflage overwhites. "It's difficult when you're going in temperatures like you are, working up a sweat. It's a bad combination. If you don't have the right gear, it's dangerous," Ito says.

He talked about the cold one day in late December 1990, while he was in his house for lunch. The cold is part of his life. He dresses for it. He eats to stay warm. A substantial part of his diet is the food he catches, high in protein and

energy: whale blubber, beluga oil, raw fish and seal oil. "It's like when you put antifreeze in your car," he says. "It keeps you from freezing."

Ito is Inupiat Eskimo; his grandmother's family is from Noatak, Kivalina and Point Hope. His Japanese grandfather and his ancestors hunted seals and whales and fished. "It's bred right into me," Ito says. "If I don't have a seal during the winter, it's hard to survive the conditions. I don't do it for a trophy."

On this day, Ito fishes for sheefish, a freshwater, sweet-tasting and somewhat oily fish that is an important subsistence food. Ito and his family use most of the sheefish he catches, some he gives away and he sells the rest. The Kotzebue area has the only commercial fishery for Alaskan sheefish. Ito usually runs two nets, November through January. He checks them daily, before and after lunch. With only two hours of light in midwinter, most of his work is done in the dark.

This is a typical midwinter day, temperatures 20 to 25 below, and the wind chill minus 60. Some days the wind chill is 100 below. In this weather, Ito drives a snowmachine an hour to his fishing spot on the frozen inlet. He uses 50-foot gillnets strung on a weighted line

beneath rows of holes drilled in the ice. There are no trees, no shelter to cut the wind. It is strenuous work, chopping ice, pulling nets, picking out the fish, resetting the nets. He covers the holes with cardboard or plywood when he finishes, and piles snow on top of the boards to add insulation. Still, the holes will freeze shut before morning. On the clearest, coldest nights, a layer of ice 5 inches thick will form.

Ito wears down-filled thermal underwear and acrylic pile pants that wick sweat away from his body. He wears air-insulated bunny boots or sealskin mukluks with hair soles. A garment crucial to his survival is a lambskin pullover, a knee-length parka with roomy sleeves that serve as a hand muff. Ito often works barehanded in the cold and wet. He has rubber gloves which get coated with ice. The gloves help cut the wind but are stiff and clumsy and get caught in the nets. So more times than not, he works without them. His hands get pretty cold. "You just keep moving,"

he says. "You try to work real fast."

The arctic cold can be mean and unforgiving. Ito recalls two brushes with death in the cold. One winter, traveling in a caravan of snowmachines from Kotzebue to Noorvik, a 40-mile trip across the Kobuk River delta, Ito was one of seven people on five snowmachines that plunged through the ice. Several confusing minutes passed. Ito was in the water, about to be pulled under by the current, when his older brother slid across the ice and pulled him out. A child about to go under was also rescued. Everyone was accounted for; all were safe though cold and wet. There was no firewood, so someone grabbed a fox-fur hat and set it afire. They broke their ice-stiffened clothes free from their bodies, knocked off the ice

and redressed, walking around to stay alive while someone went on a snow-machine for help. Two machines were lost in the water, and a couple of people froze their feet. "Everyone helped everyone else. It was pretty tough," Ito says.

Another time, Ito, his brother Abe, and a friend, John Schaeffer III, were walking 40 miles from Riley Wreck, a local landmark, to Kotzebue during breakup. They were crossing an ice-filled river when Ito went under. Schaeffer grabbed and pulled him out. They walked another few miles gathering enough driftwood for a small fire. Then Ito stripped down and wedged between the other two men. "They kept me from freezing," he says.

"Everytime you go out, it seems to be dangerous."

AUTUMN: THE MEANEST SEASON ON THE WEST COAST

No other state has as much coastline as Alaska has. In fact, all other states would have to team up to surpass the length of Alaska's coastline. Since a large part of the state borders water, it is no surprise so many of Alaska's climatic regions are coastal. While many parts of the Lower 48 need only be split into two groups, coastal and interior, Alaska's coast has many individual climates. Next stop in this exploration of climate is the west coast.

Western coastal Alaska, which borders the Bering Sea, has the state's the most varied weather patterns. This is due somewhat to the region's part-time coastline; in winter, the frozen Bering Sea treats the coast as an inland plain. As early as November, ice starts to form on the

TOP LEFT: *Two travelers battle wind chill in a walk across a frozen stretch of St. Lawrence Island. Extremely hazardous circumstances develop when wind and snow combine to create whiteouts, where no horizon is visible and ground and sky blur into one. Travelers can easily become disoriented in such conditions, lose their way, fall off cliffs or run into obstacles.* (Steve McCutcheon)

LEFT: *Tremendous fall storms have battered Nome since the town's founding nearly a century ago. A November 1913 storm damaged Front Street's famous Board of Trade Saloon.* (Courtesy of Nicki Nielsen)

RIGHT: *Advective fog, shown here rolling down a hillside at Cape Lisburne, is a type of fog caused by horizontal movement of moist air over a cold surface, and the consequent cooling of that air to below its dewpoint.* (David Roseneau)

shores of the Seward Peninsula. It begins as shorefast ice, which freezes from the land out to sea and from the water's surface down to the ocean floor. By early winter, the shorefast ice meets the floating pack ice which has migrated south from the Arctic Ocean. When the two ice families merge, the Bering Sea becomes little more than a large skating rink.

THE MAGIC OF THE POLAR SKIES

Winter in Alaska is a season of special effects, a time when light plays visual tricks in the polar atmosphere. Sunrises and sunsets stretch for hours as the winter sun skims the horizon, its long, low rays painting the landscape pink with alpine glow. Such light shows and other atmospheric displays – halos, sun dogs, glories, mirages, the green flash and the aurora borealis – have entranced arctic sky-watchers for centuries.

Polar explorers wrote of such things in their diaries, of brilliant sunsets, of ethereal glows encircling the sun and moon, of cities glimmering above the arctic ice. Eskimos learned to navigate through the frozen seas using ice blinks in the sky, and indigenous peoples throughout the North explained the aurora in any number of legends, usually linking the mysterious lights with their notions of life after death.

Of any of the optical displays in the polar skies, the aurora is by far the most studied. Once thought to be caused by refraction of light through ice particles in the atmosphere, scientists now know that the aurora results from a large-scale electrical discharge, powered by solar winds. Auroral studies are making fundamental contributions to science in the area of plasma physics and astrophysics. See *Aurora Borealis, The Amazing Northern Lights,* Vol. 6, No. 2, of

Frenchman De Mairan was among the first scientists to conclude that the aurora occurs well above the layer at which clouds form. (Shelley Schneider)

ALASKA GEOGRAPHIC® for more information on the aurora.

Other polar apparitions may pale in scientific significance to that of the aurora, but they still are fascinating and beautiful. And for the handful of scientists who study atmospheric optical effects, these natural nuances of sunlight and moonlight are more than minor esoterica; they are indicators of atmospheric conditions fundamental to large-scale air movements over the entire Earth.

"The brightness or color of the sky, the angle of the ice halo, the height of the mirage — they tell about dust in the upper atmosphere, about the crystalline form of ice in the air and about temperature structure at altitudes higher than airplanes fly," says Glenn Shaw, associate professor of geophysics at the University of Alaska Fairbanks.

But understanding the sciences behind these illusionary curiosities is not a prerequisite to appreciating their aesthetics.

Alaskans can see many of these images surprisingly often in winter. The polar atmosphere during this season frequently has the right mix of ingredients to trigger their appearance, particularly when

By far the most studied of the optical displays visible in polar skies is the aurora borealis, or northern lights. In Alaska, September and March are good months to see the aurora, although it can be seen potentially anytime when the skies are dark. (Cary Anderson)

scenes are illuminated by a low sun. Some displays are caused when ice crystals bend and reflect light; others happen when light passes through layers of hot and cold air.

Here is a quick look at weather-related anomalies such as halos, sun dogs, light pillars and mirages, and how and when to see them.

◆ Halos are circles around the sun or moon that occur when the air is full of small, six-sided ice crystals. The crystals tumble through the air, acting as prisms to refract, or bend, light as it passes through them. The most commonly observed halos are visible at 22 degrees and 46 degrees from the sun.

According to an old wives' tale, a halo around the moon signals approaching bad weather. That is a fairly dependable forecast since the ice crystals that create halos are often associated with high-altitude cirrus clouds that normally precede a storm front.

◆ Sun dogs are concentrated shields of light on either side of the sun. They are actually isolated vertical sections of the halo, created by flat, platelike ice crystals. These plate-shaped crystals fall somewhat like leaves, their flat bases oriented

horizontally. Because of the way in which flat crystals refract light, only the sides of the halo, thus sun dogs, appear to an observer on the ground. Sun dogs disappear when the sun is above about 60 degrees elevation.

◆ Light shafts or pillars are white columns of light that appear when the sun, moon or artificial lights, such as car headlights and streetlights, reflect off ice crystals. These vertical columns may appear above or below the light source. The common explanation for many years was the columns were caused by light reflecting off flat crystals falling with their bases horizontal. But work by University of Wisconsin physicist Robert Greenler has shown that light pillars can, in some

cases, come from pencil-shaped crystals spinning with their long faces horizontal.

◆ Coronas are soft, colored rings of light around the sun and moon, caused when water droplets or ice crystals in the air diffract the light. Diffraction occurs when light reflects off the back sides of a droplet or crystal. Some of the rays return

toward the light source, but in doing so create circular zones of lightness and darkness. The diameter of the zone for each color is different, creating a rainbow. The sequence of colors in a corona goes from bluish on the inside to a reddish-brown on the outside. Coronas are smaller than 22-degree halos. To see a sun corona without damage to the eyes,

LEFT: Sun dogs add a specter of otherworldliness to this image of the floatplane base at Lake Hood in Anchorage. In reality, sun dogs are concentrated shields of light on either side of the sun formed by flat, platelike ice crystals. (Jon Nickles)

RIGHT: This mirage rises over spring ice in the Chukchi Sea. Mirages take place when light passes through layers of cold and hot air, a variation that causes the light rays to bend toward the coldest layer. In the Arctic, the coldest layer is next to the ground, thus the mirage appears above the actual object. (David Roseneau)

A mirage makes a portion of the Alaska Range appear to rise from a tableland of lower mountains. Mirages seen in the Arctic are known as superior mirages, and sometimes they enable earthbound observers to see objects beyond the horizon. (Mark McDermott)

experts recommend looking at its reflection in a water puddle or window.

◆ Glories are coronas that appear around shadows that fall on clouds or fog banks. This apparition is sometimes called the "specter of the Brocken" after its frequent appearance on the Brocken, the highest peak in the Harz Mountains of eastern Germany. According to one story, a climber on the Brocken fell to his death when he was startled by the sudden appearance of a human figure in the nearby mist, its head encircled by light.

For an individual to see his own glory, his shadow must fall on a fog or cloud bank. Glories, like coronas, are caused by diffraction and thus are rainbow-colored. The glory always appears at the point opposite the sun, or the anti-solar direction, which corresponds with the head of a person's shadow.

People traveling in airplanes or helicopters can frequently see glories around the aircraft's shadow as it falls on clouds below.

Another type of halo, a *heiligenschein* or holy light, appears around a shadow that falls on the ground or on a dew-covered surface. This halo is white, rather than rainbow-colored.

◆ Ice blink and water sky are phenomena often used by Eskimos and other travelers, such as ship captains, navigating through the arctic ice pack. Because of the way light scatters through the air, the sky can mirror the ice and water beneath it.

A white ice blink appears in the sky over ice on the horizon, allowing the traveler a look at ice that may be as many as 20 or 30 miles beyond normal vision. Experienced arctic travelers say that the different types of ice "blink" with slightly different colors.

Likewise, open water in an ice-clogged ocean may reflect as dark "water sky." When combined, the ice blink and water sky can provide a detailed sky map of features on the ground. This can happen when two adjacent ground areas, such as white ice and dark water, differ dramatically in their capacity to reflect light. This difference can be seen in the brightness of the sky above, resulting in a sky map.

Robert Greenler, in his book *Rainbows, Halos, and Glories* (1980), explains how sky maps and a related phenomenon called whiteout occur. White snow and ice scatters light over and over between Earth and a thin layer of clouds. In a situation where light is repeatedly scattered between cloud and snow, no shadows exist and even the horizon can disappear. "This is a whiteout . . . where all visual cues of orientation and ground structure are gone. Lacking such cues, some people have difficulty even in maintaining their balance; travelers stumble, unseeing, into ice blocks or snow drifts or step off invisible ledges in the featureless snowscape," Greenler writes.

◆ A green flash is sometimes seen as sun rises or sets on a clear, cold day. This brilliant green light glints from the horizon as the first edge of the sun appears, or as the last edge disappears. Scientists say this happens because of atmospheric refraction and filtering. Essentially, the atmosphere acts as a prism, separating the colors of light enough to bend the red and yellow light below the rim. This leaves blue and green, but blue is scattered by the atmosphere. So the color to be seen is green.

In a paper on polar optical effects, UAF's Shaw tells of a green flash seen one December evening from the Geophysical Institute in Fairbanks when the sun was skimming the tops of the Alaska Range, 90 miles away. The emerald green light lasted about four seconds until it was blocked by a mountain peak. The green light then reappeared for another three

Saucer-shaped lenticular clouds indicating high winds at altitude form over Cape Lisburne north of Point Hope on the Chukchi Sea coast. (David Roseneau)

A low-angled sun spreads alpenglow, probably the most common manifestation of the magic of polar skies, across this snowy vista in the Copper River valley. (Ruth Fairall)

their image above the horizon.

For starters, mirages occur when light passes through layers of hot and cold air. Light bends toward the densest, or coldest, layer of air. In the desert, air nearest the ground is the warmest so light passing near the Earth's surface will bend upward toward the cooler air. This produces an inverted image below the real object, called an inferior image.

Arctic mirages, labeled superior mirages, appear over the object because the air nearest the ground is cooler and thicker than the warmer air a few feet higher, thus creating conditions for light to bend toward the Earth. This is also known as looming and can allow sight of an object otherwise below the horizon.

A complex arctic mirage, the fata morgana, occurs with temperature inversions, where there may be several layers of warm and cold air. The fata morgana may project a double image of an object, or be more complex, creating phantom cities complete with towers and spires.

seconds as the sun came out from behind the peak. "The light intensity was strong enough to cast distinctly verdant hues on the landscape," he writes. "It would have been an appropriate show for St. Patrick's day."

Greenler says if the sun near the horizon appears very red, do not expect to see the green flash. But if the sun remains yellow as it sets, the flash is likely to be visible.

◆ Mirages are often thought of as desert apparitions, such as a shimmering oasis beckoning a parched traveler. But mirages also appear in the Arctic, where some of the most elaborate mirages occur. Because of arctic mirages, early explorers charted islands and mountains where there were none, creating great confusion among subsequent expeditions. Some mirages may have actually contributed to the discovery of lands by projecting

The name comes from Celtic legend and Arthurian romance. Morgana was a fairy, a half-sister of King Arthur. She was an enchantress, said to live in a crystal palace beneath the sea. She was skilled in the art of mirage, having learned her powers from Merlin the Magician. She sometimes made her castle appear above the waves. Seamen mistaking it for safe harbor were lured to their death, so the legends go. Fata morgana means Morgan le Fay, or Morgan the fairy, and is a centuries-old name for complicated mirages.

A famous fata morgana arose in southeast Alaska, from Muir Glacier in the late 1800s. This "Silent City" was seen several times, according to accounts in *The Wonders of Alaska Illustrated*, by Alexander Badlam (1890). Badlam wrote of seeing the city while cruising through Glacier Bay aboard the steamship *Ancon*. The vision included 300-foot-tall spires above buildings with doors and windows, and amid streets and gardens. A later account from another group said the city was of extensive proportions with houses, churches and perhaps 15,000 or 20,000 inhabitants. These sightings lead to one of the biggest hoaxes of the time. Dick

Willoughby, a Juneau prospector who had a colorful history of money-making gimmicks, started taking visitors on tours of Glacier Bay to see the Silent City. Shortly thereafter, he produced a photograph of the mystical city and sold copies of it for 25 cents, 50 cents, then 75 cents each to the eagerly awaiting tourists. A scientist eventually pronounced the photo a hoax, and the city was identified as Bristol, England. Willoughby had purchased the camera and its plates for $10 from an English photographer stranded in Juneau. Even after the deception was revealed, tourists continued buying the Silent City photo and going on his tours, still hoping to see the city in the sky. In 1900, two years before his death, Willoughby sold the negative for $500.

Layers of air of varying temperatures create this mirage of the Alaska Range as seen from Fairbanks with the Tanana Valley in the foreground. (Glenn E. Shaw)

Cooling Power of Wind

(Expressed as Equivalent Wind Chill Temperature)

WIND SPEED (IN MPH)	TEMPERATURE (°F)																				
CALM	40	35	30	25	20	15	10	5	0	-5	-10	-15	-20	-25	-30	-35	-40	-45	-50	-55	-60
5	35	30	25	20	15	10	5	0	-5	-10	-15	-20	-25	-30	-35	-40	-45	-50	-55	-65	-70
10	30	20	15	10	5	0	-10	-15	-20	-25	-35	-40	-45	-50	-60	-65	-70	-75	-80	-90	-95
15	25	15	10	0	-5	-10	-20	-25	-30	-40	-45	-50	-60	-65	-70	-80	-85	-90	-100	-105	-110
20	20	10	5	0	-10	-15	-25	-30	-35	-45	-50	-60	-65	-75	-80	-85	-95	-100	-110	-115	-120
25	15	10	0	-5	-15	-20	-30	-35	-45	-50	-60	-65	-75	-80	-90	-95	-105	-110	-120	-125	-135
30	10	5	0	-10	-20	-25	-30	-40	-50	-55	-65	-70	-80	-85	-95	-100	-110	-115	-125	-130	-140
35	10	5	-5	-10	-20	-30	-35	-40	-50	-60	-65	-75	-80	-90	-100	-105	-115	-120	-130	-135	-145
40	10	0	-5	-15	-20	-30	-35	-45	-55	-60	-70	-75	-85	-95	-100	-110	-115	-125	-130	-140	-150

(winds above 40 mph have little additional effect)	Little danger	Increasing danger (flesh may freeze within 1 minute)	Great danger (flesh may freeze within 30 seconds)

DANGER OF FREEZING EXPOSED FLESH FOR PROPERLY CLOTHED PERSONS

Source: Alaska Regional Profiles, Yukon Region; *graphic by Kathy Doogan*

The weather at Nome offers a good example of the thermal result of this cycle. The change of climate is evident in the town's average temperatures: Extremely cold temperatures from December through March reflect the loss of moderating effects from open water. Breakup of the ocean ice contributes to rapid temperature increases during April and May. By May's end, the ice is gone, and the ocean is back to its temperature-taming ways.

Winter is bitterly cold and unkind to the west coast. Strong westerly winds, no longer tempered by open water, make temperatures fall well below zero. Wind-chill factors frequently reach equivalent temperatures of 50 below zero. Thanks to the frozen sea, the west coast experiences all the disadvantages of a continental winter climate.

Summer, though, is a much different story. The open water of the Bering Sea moderates the region's temperatures, just as the Gulf of Alaska does those of the panhandle. West coast summers are slightly cooler than those of Southeast because the prevailing westerly winds carry moisture and cooler air from the sea.

In most areas, summer and winter are the seasons of extremes. Along the west coast, however, spring and fall claim that distinction. These are the seasons of transition, when the region feels the force of changing weather patterns.

Spring along the Bering Sea coast means floods, as frozen rivers from as far away as Yukon Territory send their melted ice westward.

The bulk of this ice makes it way to the sea through the delta of two of Alaska's largest rivers, the Yukon and the Kuskokwim. Spring breakup also means thawing of southwestern Alaska's marshy plains, and the sudden release of moisture into the still-chilly air. This creates the thick fog synonymous with spring in this region. Autumn is especially violent along the west coast, mostly due to the first storms of the winter season. Throughout the year, strong, low-pressure storm systems move from west to east. These storms usually follow the Aleutian Islands until they reach Alaska's mainland. From there, the storms take either of two paths, known as storm tracks. The southern route, across the Gulf of Alaska, the panhandle and eventually into Canada, is the most common path during the summer. Winter favors the northern route, where storms turn left at Bristol Bay, and proceed northward through Bering Strait. As autumn wears on, more and more storms opt for the northern route.

Storms moving along the northern track are still strong when they hit the west coast. The wind can blow more than 50 knots from the same direction for days at a time. This consistently strong wind creates storm surge floods, the result of tremendous masses of ocean water having been pushed ashore by the wind. Nome has recorded several memorable fall storms that have sent waves surging over Front Street, washing away buildings facing Norton Sound. With all the potential for hazardous weather during autumn, the west coast may be the only part of Alaska where winter is not the meanest season.

After digging down through almost six feet of snow on Judd Lake to locate and tie up a piece of dock that had blown away during a late October storm, David Fox stops to peer out at his cat, Gracie Lynn. If he did not tie the piece of dock to a tree, it would float away down Talachulitna Creek during the high water of spring breakup. (Shelley Schneider)

THOMPSON PASS SNOW BUSTERS

FOURTEEN-FOOT POLES LINE EACH EDGE OF the Richardson Highway through Thompson Pass near Valdez. In summer, when the tundra of the pass is mostly green with patches of snow, these giant metal probes seem without purpose. But come winter, it is a different story.

Thompson Pass catches some of the most severe winter driving conditions of any place in Alaska. It holds four state snowfall records: the greatest snowfall in 24 hours, 62 inches in December 1955; the greatest snowfall in one storm, 175.4 inches, also in December 1955; the greatest snowfall in one calendar month, 297.9 inches in February 1953; and the greatest snowfall in one season, 974.5 inches in 1952-53. The average snowfall from 1952 to 1987 has been 485 inches, with exceptionally heavy snowfalls of 750 inches and more the winters of 1987, 1988 and 1990. The winter of 1989 was routine, about 450 inches. Winds at the top of the pass average 48 knots, and gusts routinely reach 100 knots. Drifts can form quickly, and driving becomes hazardous.

These poles help motorists find their way when blowing snow obliterates most everything else. At the top of each road marker is a four-foot arm extending over the highway, and on each arm is a length of yellow reflector tape. The tape is 36 inches long on the arms across the lane in which the driver should be; the tape is half that length on the arms across the opposite lane. "Twenty times a winter that is important, to give someone a bearing of which side of the road

Snow removal equipment tackles some of the many feet of snow that annually falls on Thompson Pass near Valdez. The pass area claims all of the snowfall records for the state, with a record of 974.5 inches in 1952-53. (Courtesy of Alaska Department of Transportation and Public Facilities)

Winds of 100 mph sweep across this field at Valdez in January 1966. High winds are just one element of weather that observers record. (Pete Martin)

they're on," says Jim Britt, foreman of the state highway camp for the pass.

Britt has been working in Thompson Pass for 22 years. He and a 10-person wintertime crew rotate shifts to keep snowplows, blowers, graders, bulldozers, front-end loaders and sanders working seven days a week, night and day. From September to May, they maintain a 24-mile stretch of road through the pass, mile 18 to mile 42. The Richardson Highway is the only overland route in and out of Valdez.

In January 1989, a winter storm closed the highway through the pass for three days. Record cold temperatures were recorded in 33 weather stations throughout Alaska, and winds raged in the pass for seven days. It was the worst stretch of weather Britt has experienced here. Winds up to 140 mph whipped deep snow on the ground into drifts 19 feet high across some stretches of the road, Britt recalls.

The wind also blew over a mail truck and a grocery truck which had stopped because of poor visibility. All told, the storm stranded 26 travelers in the pass.

The road crew brought them from their cars to the highway camp's bunkhouse, a two-story, 1940s-vintage building with a kitchen, dining room, laundry room and more than a dozen bedrooms. For four days, the rescued motorists and the road crew shared quarters and an emergency supply of frozen TV dinners. "I don't think I've eaten one since," quips Britt.

Storms strand motorists several times every winter, but usually only for a day or overnight. What often happens when a storm hits the pass, says Britt, is that drivers are blinded by blowing snow in areas where the wind is worst. "They go through one or two whiteout conditions, then come up to another one and think

it's the same and end up hitting a drift or another car. If they'd take their time going through, 99 percent would make it," Britt says. "If you can't see, stop. That's what we do.

"If we can keep traffic moving, it's easy. When we have to start pulling cars out of ditches, we get behind and end up with trouble."

But maybe the hardest job for the Thompson Pass snow busters is waiting for spring to come to the highlands of the Chugach Mountains. "Down below, the snows are melting and the temperatures are warming, and we're still fighting snow up here," Britt says. "Everyone gets antsy for summer."

THE ARCTIC: FAR FROM THE DISTANT SUN

The climate most Outsiders associate with Alaska resembles that of the Arctic. Along the North Slope, that area extending from the northern foothills of the Brooks Range to the Arctic Ocean, the climate is influenced by ice floes and the midnight sun. While this part of Alaska has the greatest fluctuation in daylight, it does not experience a similarly wide fluctuation in temperatures

In winter, Barrow's below-zero average monthly temperatures are due to the small amounts of, or total absence of, sunlight from October to March. With no solar energy to offset the constant wintertime loss of heat to the atmosphere, Barrow's average temperatures fall well below zero, and remain there until May. The only protection from energy loss comes from thick blankets of clouds or fog, or the occasional arrival of warm air masses, courtesy of powerful winter storms and their fronts.

While winter along the arctic coast is a frigid affair, mostly because of the sun's absence, summer does not bring the opposite. In the first place, the Arctic Ocean, frozen clear to the sea floor for miles from shore in winter, is free of ice in summer. The water moderates temperatures here just as it does along the west coast. The state's northernmost coast has yet another solar phenomenon at work in summer: Even though this season offers continuous daylight, the quality of the energy from the non-setting sun is poor. The sun's angle above the horizon is low, forcing solar radiation to penetrate a thick layer of atmosphere. Because it strikes the surface at such an oblique angle, each

ray of weakened sunlight is spread across a large area of the Earth's surface. This results in a relatively low concentration of weak sunlight, and despite its 24-hour duration, this weak energy simply cannot boost temperatures much.

The North Slope is normally spared frequent bouts of severe weather, mainly because this part of Alaska lies at the end of the northern storm track. Many low pressure storms run out of energy by the time they reach here. Often, the region's most punishing weather comes from storms born in the Arctic Ocean or the Soviet Union, storms still relatively young and strong when they reach the slope.

Noctilucent Clouds: Little-Known Denizens of the Polar Skies

THE POLAR SKIES CREATE A NUMBER OF intriguing and beautiful phenomena. Among these are noctilucent clouds, pearl white waves of tiny ice crystals that coalesce in the atmosphere's upper limits, miles beyond where clouds normally are found.

Noctilucent clouds not only form where clouds should not, they are relatively young to science, appearing for the first time only a century ago. In the past few decades, however, the number of noctilucent clouds has increased significantly, and atmospheric scientists today think that these high-floating apparitions are evidence of global warming.

Noctilucent clouds are unique to subarctic latitudes in the northern and southern hemispheres, and show up best during the six weeks following summer solstice.

Neal Brown, assistant professor of physics at the University of Alaska Fairbanks, has studied this type of cloud

Noctilucent clouds, white waves of ice crystals unique to subpolar latitudes in the northern and southern hemispheres, develop high above where other clouds form and are so immense that one cloud may cover the entire state of Alaska. (Neal Brown, Geophysical Institute, University of Alaska)

off and on since 1964, when he joined a noctilucent cloud project at the university's Geophysical Institute. He spent the following 18 years as director of the Poker Flat Research Range, primarily concerned with studying the aurora borealis, but he never lost his fascination with noctilucents. In summer 1990, he took his first vacation in two decades away from Fairbanks and visited noctilucent cloud researchers at the University of Colorado's Laboratory for Atmospheric and Space Physics in Boulder.

"Noctilucent clouds are a phenomena that's just beautiful," Brown says enthusiastically. "They're the equivalent of bowling balls hanging in the air right in front of you. They defy our theories to explain how they can form and be there."

Noctilucent clouds occur some 50 miles above the Earth. That is high considering that most clouds form within seven miles of the ground. Noctilucent clouds form in the mesosphere, which starts about 30 miles above the Earth and extends to an altitude of about 53 miles. Air is exceedingly thin in the mesosphere. For instance, 90 percent of the Earth's air is below the top of Mount McKinley, which at 20,306 feet is less than four miles high. Of the remaining air, 90 percent is found in the next five miles. So by the time the mesosphere is reached, explains Brown, little air remains. Scientists puzzle over how this thin atmosphere could contain enough water vapor and dust to make clouds, much less those the size of noctilucent clouds. It is not unusual for these clouds to cover an area greater than the reach of Alaska, which measures 586,412 square miles. A typical noctilucent cloud will cover the sky in an area bordered by the coasts of Alaska and Canada on the north and Kodiak and Hudson Bay on the south.

The discovery of noctilucent clouds came about after the Krakatoa volcano in Indonesia erupted in 1883. The explosion spewed tons of water vapor and dust into the air, blackening the atmosphere of the northern hemisphere for almost three years. In 1885, after Krakatoa's fallout had migrated into the mesosphere, the first noctilucent cloud was sighted over Germany. Within a few days, astronomers and meteorologists throughout Europe reported the phenomenon.

The increase of noctilucent clouds in recent decades, however, is linked to man's activities rather than nature's. University of Colorado researchers think that increasing amounts of methane in the atmosphere are triggering an increase in these mesospheric clouds. Man's activities have nearly doubled the amount of methane in the environment; air bubbles trapped in polar ice show

The cloud that frequently hangs over but does not touch Mount McKinley is often irridescent, and since it is formed in the same way nacreous clouds are formed, by a land mass pushing air into colder, higher altitudes, this could be considered a nacreous cloud. However, this cloud forms too low to be a true nacreous cloud, and may be more properly called a lenticular cloud. (Shelley Schneider)

much lower concentrations of methane before the industrial age. Methane is a greenhouse gas, more effective at trapping heat than carbon dioxide, the most commonly named culprit in the theory of global warming. As methane moves up through the atmosphere, it breaks down and forms, among other things, water vapor. As long as this is happening, scientists expect the number and brightness of noctilucent clouds to continue increasing.

Noctilucent clouds can be seen on summer evenings, after the sun has set. Noctilucent, in fact, means night shining. Because these clouds form so high above the Earth, they can be seen only when the sun illuminates their bottoms. This happens a half-hour to hour after the sun has disappeared below the horizon, and the lower atmosphere is dark. They are always white, sometimes lacy in appearance, sometimes appearing as crests of waves. Sometimes they appear as two sets of waves coming from different directions. They move much slower than lower clouds.

Sometimes nacreous clouds are mistaken for noctilucent clouds, says Brown, but nacreous clouds are rainbow-

colored while noctilucent clouds are white. Because they are white, noctilucent clouds are thought to contain tiny ice particles; larger ice crystals of various sizes result in iridescence, like that seen in nacreous clouds. Nacreous clouds are considered the finest and most colorful of any iridescent cloud, which includes various mother-of-pearl displays such as sun dogs. Nacreous clouds form 12 miles to 20 miles above the Earth, the second highest clouds after noctilucents. However, nacreous clouds are caused by a structure on Earth, such as mountains, that push air into the higher, colder altitudes. Nacreous clouds are visible only when the sun is low or actually beyond the horizon, but they disappear by the time the sun is low enough to illuminate noctilucent clouds. The cloud sometimes seen hanging above Mount McKinley, like a cap but not touching the mountain, is formed in the same way as nacreous clouds but is usually considered a lenticular cloud because it develops lower in the atmosphere.

Noctilucent clouds are best seen in Alaska from latitudes 52 degrees north to 70 degrees north. Glennallen is a prime viewing spot, says Brown.

ALASKA'S WEATHER RECORDS

Source: NOAA Technical Memorandum NWS WR28, *Weather Extremes*, Robert J. Schmidli, Weather Service Forecast Office, Phoenix, Ariz., Western Region, Salt Lake City, Utah. Nov. 1971 (updated by Jim Wise, state climatologist, Alaska Climate Center).

Graphic by Kathy Doogan

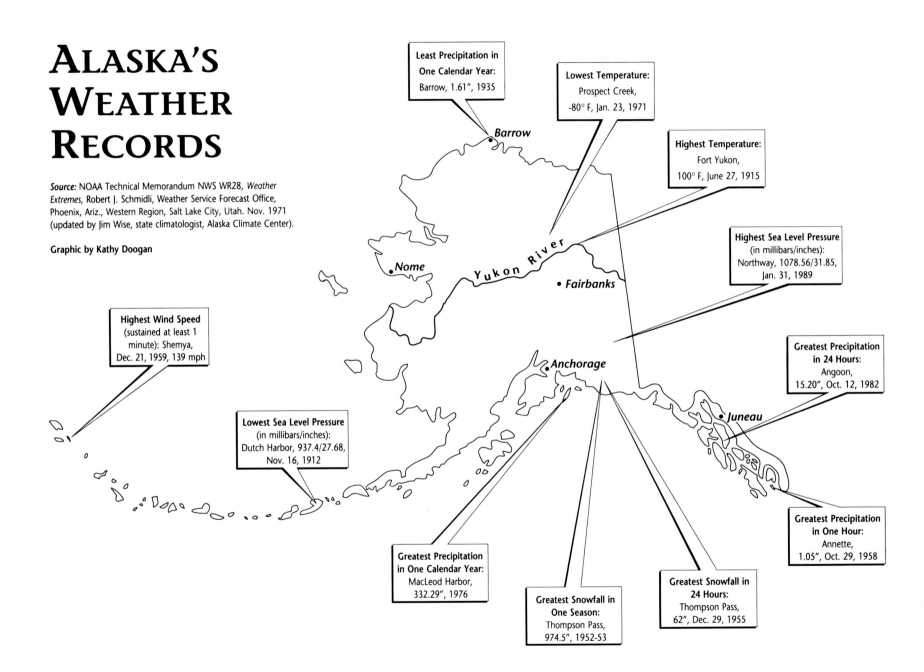

Least Precipitation in One Calendar Year: Barrow, 1.61", 1935

Lowest Temperature: Prospect Creek, -80° F, Jan. 23, 1971

Highest Temperature: Fort Yukon, 100° F, June 27, 1915

Highest Sea Level Pressure (in millibars/inches): Northway, 1078.56/31.85, Jan. 31, 1989

Highest Wind Speed (sustained at least 1 minute): Shemya, Dec. 21, 1959, 139 mph

Greatest Precipitation in 24 Hours: Angoon, 15.20", Oct. 12, 1982

Lowest Sea Level Pressure (in millibars/inches): Dutch Harbor, 937.4/27.68, Nov. 16, 1912

Greatest Precipitation in One Hour: Annette, 1.05", Oct. 29, 1958

Greatest Precipitation in One Calendar Year: MacLeod Harbor, 332.29", 1976

Greatest Snowfall in One Season: Thompson Pass, 974.5", 1952-53

Greatest Snowfall in 24 Hours: Thompson Pass, 62", Dec. 29, 1955

Barrow

Nome

Yukon River

• Fairbanks

Anchorage

• Juneau

THE INTERIOR: LAND OF EXTREMES

If the North Slope is the climatic region everyone associates with Alaska, then the Interior is the region about which everyone is warned. This is the land of extremes. It is Alaska at its most beautiful, and most dangerous. Most of the state's records for wind and precipitation were set in coastal locations where there are no barriers to slow wind approaching from the sea and where the ocean is an endless source of precipitation. But the temperature records belong to the Interior where the state's lowest recorded temperature, 80 degrees below zero, was registered at Prospect Creek on January 23, 1971. Interior residents claim to have recorded unofficial temperatures well below this record.

However, the Interior's temperature records are not all on the low side of the freezing mark. Fort Yukon, one of Alaska's coldest spots in winter, also claims the only official 100-degree temperature ever recorded in the state, on June 27, 1915. The same set of environmental circumstances that generate the extreme lows also produce the highs.

Throughout this region, temperatures follow the sun. During winter's long nights, temperatures routinely fall to levels colder than even those of the Arctic. Summer's long days generate the warmest average temperatures in Alaska. Fairbanks' records offer a reasonable example of the temperature spread found in interior Alaska. The difference between the city's average temperatures for July and January is more than 70 degrees; temperature ranges this wide are quite common in the Interior.

The favorite season for many Alaskans, fall brings crisp, insect-free air and a kalidescope of colors to the countryside such as these hills of the Yukon-Tanana Uplands. (Charles Kay)

There are no meteorological conditions peculiar to the Interior that create such temperature fluctuations. Rather, it is the absence of the temperature-moderating conditions found along the coast that allows such extreme variations. Since dry air is the most thermally active, and since the driest air in the state is found in its midsection, the Interior's location away from the oceans accounts for the tremendous fluctuations in temperature.

Weather in the Interior often resembles that in the central part of the Lower 48, where cumulus clouds, thunderstorms and even an occasional tornado occur. Once again, it is the distance from water which allows these events to happen. Thunderstorms require the rapid rising of heated air. The dry air of the Interior heats quickly under the summer sun, and easily rises into convective cells. The region's many valleys lack the mountainous terrain which disrupts convection. In some locations, convective showers deposit enough summer rain to give these places higher yearly rainfall than many coastal sites.

This brief tour of Alaska's climate may answer some questions about how the state's weather operates. But it may also ask new ones, inspiring further investigation into the how's and why's of weather on the Last Frontier. But whether the interest in Alaska's climate is comprehensive or casual, looking at the actual climate data, as well as the physics behind the numbers, will always be the most effective way to find the answers.

A wind storm sends snow flying across the highway at Big Delta in the Tanana River valley. (Pete Martin)

MARINE PILOT JACK JOHNSON

MARINE PILOT JACK JOHNSON, 64, HAS been tossed about, iced over, fogged in and amazed by Alaska's weather since age 13, when he first went to sea as a deck boy. He has fished and crabbed in

Marine pilot Jack Johnson, 64, says bad weather at sea "is just a part of life." Johnson should know. He has been at sea since he was a 13-year-old deck boy. (Gary Daily)

waters off Kodiak, served on mail boats in the Aleutian Islands, piloted state ferries and today makes his living taking ships in and out of Alaska's deep-water harbors.

Johnson has experienced most every type of marine weather imaginable, from hurricane force winds and freezing temperatures that ice decks with ocean spray to fog so thick that visibility is nil. "I'm always impressed with the weather," Johnson says, "especially when it's flat and calm, a beautiful sunshiny day with birds singing."

This January day, however, is not one of those times. The wind is blowing at a steady 45 knots, with gusts to 70 knots. Johnson is waiting out the wind, sipping coffee inside the harbor master's office at Dutch Harbor, Unalaska. The Aleutian Islands are known as the birthplace of the winds because of a parade of low pressure systems that sweep overhead. Winds commonly blow in excess of 75 knots on the Bering Sea. Sometimes in Dutch Harbor winds blow more than 120 knots.

Johnson, one of 22 marine pilots in the Southwest Alaska Pilots Association, works in Dutch Harbor frequently, piloting ships from the open ocean to the dock. In this job, he is accustomed to weather holds of 24 to 48 hours while the winds subside. Plenty of other times, though, he has been on the seas in high winds. "It's rather awesome," he says. "You feel a strong, strong, almost electric feeling in the air. The seas keep getting higher and higher. There's the screaming of the wind. Oh God, your sea is so large and my vessel so small. . . ."

This day, waiting for Johnson in the waters outside the harbor is the S.S. *President Tyler*, a massive container ship enroute from the west coast of the United States to Japan and other points in the Far East. The ship will dock in Dutch Harbor to take on additional cargo. At a length of 800 feet — more than two-and-a-half football fields — and with a deep

A crewman chips ice from the deck of an American trawler at Dutch Harbor. Chipping ice can be a tedious chore. "Sometimes it seem[s] a lot easier to drown than break the ice off," says Jack Johnson. (Gary Daily)

draft of 35 feet, 6 inches, conditions have to be just right to safely bring in the ship.

"The direction of the wind has a lot to do with it," Johnson says. "A steady 40-45 knots would push the ship toward the dock and if anything were to happen, it could hit the dock. Smaller vessels we could bring in during this weather, but it's not comfortable."

Perhaps the worst seafaring conditions, though, are high winds and frigid temperatures together, as often occurs in Alaska's winter seas. The winds whip the waves overboard, sending sheets of water over the deck and pilot house. The water and spray instantly freezes, coating everything on deck in thick ice layers. Icing makes the ship top-heavy, a particularly precarious state in wind. It is not an uncommon occurrence; such weather is one of several

things that contributes to making fishing Alaska's waters one of the most dangerous occupations in the nation.

Johnson recalls a trip in winter 1955 when he was on the mail boat *Expansion* bound for Port Graham out of Kodiak. The boat iced up and began listing. By the time they reached shelter in the Barren Islands, the boat was laid over some 30 degrees. "We chipped ice all night. Sometimes it seemed a lot easier to drown than break the ice off," Johnson says. "It can get a little hairy. You can become a little nervous."

Given the nature of Alaska's weather at sea, one might wonder why Johnson, a grandfather and great-grandfather, would return day after day to its peril. Partly, he says, it's in his blood; his father, "a tough old Finn," spent his life at sea. Besides, the bad weather is just part of the package. "It happens, it's over. It's just a way of life," he says.

Once, in 1988, when Johnson was the Alaska pilot for an international cruise

around the northern coast of Alaska through the Northwest Passage, the ship got stranded for five days in the ice pack of the Beaufort Sea. A dozen polar bears gathered around the stern while the crew worked to free the ship before the ice crushed it.

Sometimes, the weather makes for memorable experiences of the pleasant kind. Another year, 1985, as the Alaska pilot on the same Northwest Passage route, Johnson witnessed an unusual sight. He saw a vivid occurrence of a meteorological phenomena called the green flash.

A green flash is easiest to see on cold, clear days as the sun is just rising or setting. The atmosphere acts like a weak prism to bend the colors of light and under certain conditions, green is the only color visible, usually for only a split second.

The green flash Johnson saw on this summer eve lasted some 15 minutes, he said, as the edge of the sun traveled along the horizon between sunset and sunrise. "I've seen the green flash many times at sea," Johnson says, "but this was so exceptional that the captain and I both logged it. It was a once in a lifetime — once in several lifetimes — sighting." ◆

BIBLIOGRAPHY

Corliss, William R. *Rare Halos, Mirages, Anomalous Rainbows And Related Electromagnetic Phenomena*. Glen Arm, Maryland: The Sourcebook Project, 1984.

Davis, Neil. *Alaska Science Nuggets*. Fairbanks: University of Alaska Geophysical Institute, 1984.

Greenler, Robert. *Rainbows, Halos, and Glories*. Cambridge, England: Cambridge University Press, 1980.

Helfferich, Carla. "Noctilucent Clouds Signal Pollution." Ketchikan: *Southeastern Log*, August 1989. From weekly *Alaska Science Forum* column supplied by University of Alaska Geophysical Institute, Fairbanks.

Lowry, Shannon. "Safe Port In A Storm." *Alaska* magazine, November 1988.

McKinney, Debra. "Arctic Ghosts." *Anchorage Daily News*, January 25, 1990.

Shaw, Glenn E. "Observations and Theoretical Reconstruction of the Green Flash." First appeared in *Pure and Applied Geophysics* 102: 223-235. 1973.

—. "Optics of Polar Atmospheres." First appeared in 1978-1979 *Geophysical Institute Annual Report* 1-14. Fairbanks: University of Alaska, May 1980.

Thomas, Gary E.; Olivero, John J.; Jensen, Eric J.; Schroeder, Wilfred and Toon, Owen B. "Relation between increasing methane and the presence of ice clouds at the mesopause." *Nature,* Vol. 338: 490-492. April 1989.

INDEX

A

Admiralty Island 19
Alaska Range 37, 56
Alaska's maritime regions 32
Aleutian Islands 32, 73
Alexander's Dark Band 40
Alpenglow 58
Amchitka Island 7
Anchorage 11, 25, 29, 55
Ancon 59
Angoon 17
Annette 17, 34, 43
Anti-solar point 38, 40
Arctic, the 64
Arctic Village 64
Atmospheric displays 52
Aurora borealis 52, 53

B

Bald eagles 25
Baranof Island 18
Barrow 31, 64
Bikulcs, Lucy 20
Breakup 61
Britt, Jim 63
Bubbel, Bill 31
Budke, Bob and Polly 17

C

Cape Lisburne 57
Chilkat River 25
Chitina River valley 39
Chukchi Sea 55
Circulation model 22
Climate 11, 12
Climatic regions 14
Clouds, lenticular 57, 69
Clouds, nacreous 67, 69
Clouds, noctilucent 66-69
Cold, state record 71
Copper River valley 58
Coronas 55, 56

D

De Mairan 52
Direct sunlight 14
Dozier, Debra 18-20
Dutch Harbor 73, 74

E

Eidler, Jane 20
Expansion (mail boat) 74

F

Fairbanks 31, 44, 45, 71
Fata morgana 59

Fog, advective 50
Fort Yukon 45, 71
Frozen bodies of water 29
Fur Rendezvous 7

G

General circulation 23
Glennallen 69
Glories 52, 56
Green flash 52, 57, 58, 74
Greenhouse gas 69
Gulls 25

H

Haines 17, 25
Halos 52, 53
Heat, state record 45, 71
Heiligenschein 56
High pressure systems 23
Homer 33
Hubbard Glacier 40
Hulahula River 7, 36

I

Ice blinks 52, 56
Ice fishing 48, 49
Icy Cape 27
Inland plain 50

Interior, the 71
Ito, Gordon 48, 49

J

Johnson, Jack 73, 743
Judd Lake 12, 42, 61
Juneau 18, 21, 33

K

Katmai National Park 38
Kenai Peninsula 42
Ketchikan 17
Ketchikan tennis shoes 16
Kirchhoff, Mark 20, 21
Kodiak 32, 67, 73, 74
Kotzebue 48
Kovolisky, Joe 8
Krakatoa volcano 69

L

Lake Clark National Park 28
Light shafts 55
Little Port Walter 17
Low pressure systems 22, 23

M

Mesosphere 69

PHOTOGRAPHERS

ALASKA GEOGRAPHIC® back issues

NEXT ISSUE:

Alaska's Volcanoes: Northern Link in the Ring of Fire, Vol. 18, No. 2. More than 15 years ago, the Society published its first issue on the state's mountains of thunder. This issue will bring readers up to date on Alaska's living mountains. To members 1991, with index. $17.95.

ALL PRICES SUBJECT TO CHANGE.

Your $39 membership in The Alaska Geographic Society includes four subsequent issues of *ALASKA GEOGRAPHIC®*, the Society's official quarterly. Please add $4 for non-U.S. memberships.

Additional membership information is available upon request. Single copies of the *ALASKA GEOGRAPHIC®* back issues are also available. When ordering, please make payments in U.S. funds and add $1.50 postage/handling per copy. Non-U.S. postage extra. To order back issues send your check or money order and volumes desired to:

The Alaska Geographic Society

P.O. Box 93370, Anchorage, AK 99509

BACK IN PRINT!

13 of our most popular ALASKA GEOGRAPHIC®s

THE SILVER YEARS, VOL. 3 NO. 4 ... A pictorial history of the glory years of the early Alaska canned salmon industry. 168 pages, $17.95

SOUTHEAST: ALASKA'S PANHANDLE, VOL. 5 NO. 2 ... Our most popular issue, that we just can't seem to keep in stock. Fold-out map, 192 pages, $19.95

ALASKA WHALES AND WHALING, VOL. 5 NO. 4 ... Great photos and poster showing whale species in Alaska ... includes a fascinating history of the Yankee whaling fleet. 144 pages, $19.95

AURORA BOREALIS, VOL. 6 NO. 2 ... Glorious photos and masterful text by the famous aurora authority Dr. S.-I. Akasofu. 96 pages, $14.95

ALASKA'S GREAT INTERIOR, VOL. 7 NO. 1 ... Land of the Athabascans, Denali, the Tanana, the Yukon ... Alaska's "golden heartland." Fold-out map, 126 pages, $17.95

THE ALEUTIANS, VOL. 7 NO. 3 ... Home of the Aleuts and smoking mountains ... includes the story of the "thousand mile war." Fold-out map, 224 pages, $19.95

ALASKA'S GLACIERS, VOL. 9 NO. 1 ... Examines in depth more than a dozen glacial regions in Alaska. Great photographs and glossary ... written by Bruce Molnia of the U.S. Geological Survey. 144 pages, $19.95

SITKA AND ITS OCEAN/ISLAND WORLD, VOL. 9 NO. 2 ... The story of Sitka when it was capital of Russian America and principal city of the eastern Pacific. Includes an update on Sitka today, a bustling fish, logging and tourist center. Fold-out map, 128 pages, $19.95

ALASKA NATIVE ARTS AND CRAFTS, VOL. 12 NO. 3 ... A beautiful catalog of ancient crafts and modern: ivory, jade, masks, beaded moosehide, silver and gold ... a lovely volume. 216 pages, $17.95

ALASKA'S SEWARD PENINSULA, VOL. 14 NO. 3 ... Where Wyatt Earp and Rex Beach became names to remember and Eskimos live by the bounty of the seas and reindeer herds ... includes Nome, Golovin, Wales and more. Fold-out map, 112 pages, $15.95

GLACIER BAY, VOL. 15 NO. 1 ... The northernmost national park in Southeastern Alaska, where cruise ship visitors see harbor seals, humpback whales and orcas, and glaciers calving into the sea. One of the wonders of the world. Fold-out map, 104 pages, $16.95

DENALI, VOL. 15 NO. 3 ... The highest peak in North America, "The Great One" of Indian lore. Includes the wild denizens of Denali National Park. Fold-out map, 96 pages, $16.95

THE NUSHAGAK RIVER, VOL. 17 NO. 1 ... A great river that drains a vast interior that is beaver country and spawning grounds for hordes of salmon coming from Bristol Bay. Includes the nation's largest state park, Wood-Tikchik. Fold-out map, 80 pages, $17.95

THESE TITLES REPRESENT A LIMITED PRINTING ... ORDER NOW!

What is behind polar bear-human encounters . . . see page 86

ALASKA'S WETLANDS: MIRED IN CONTROVERSY

BY L.J. CAMPBELL

Cottongrass is found in wetlands throughout much of the state. Tufts of down from this plant were spun into fiber or used as candlewicks in earlier times. *(Chlaus Lotscher)*

IN COMING MONTHS, CONGRESS will tackle the thorny question of how to save the nation's remaining wetlands.

This goal, known as "no net loss" of wetlands, has triggered discussion throughout the United States. But few places is the debate as convoluted as in Alaska, which has more wetlands — marshes, bogs, wet meadows, mud flats — than the rest of the nation combined. In addition, Alaska is the only state with permafrost wetlands, a seemingly technical distinction but one with dollars behind it. Permafrost wetlands make up nearly all the arctic coastal plain, the nation's primary oil field and Alaska's biggest contributor to the state budget.

Both simple and complex, the wetlands question is one of conservation versus development, of soggy land, endangered birds, water use, subsistence, forests, fish and oil. It has international interest as well, with more than 60 countries looking to the United States for guidance in wetlands preservation.

At issue is President George Bush's promise to stop wetlands disappearance. "I believe this should be our national goal, no net loss of wetlands," Bush said during a 1988 campaign speech in Michigan.

A cabinet-level task force on wetlands is looking at ways to do this, in anticipation of the topic coming before Congress. Wetlands are covered by the Clean Water Act, which comes up for revision in 1992. Congress also may consider passage of new stand-alone wetlands regulations.

During 1990, the Domestic Policy Council Task Force on Wetlands held public hearings in six towns across the United States, including Anchorage.

They are also reviewing existing regulatory and non-regulatory programs to develop a list of wetlands management options. In separate action, four agencies are in the final stages of reviewing changes to the federal manual that defines wetlands, and sometime during the process will be looking at the uniqueness of Alaska's wetlands, said Ed Goldstein, the task force's senior policy analyst.

Even though "no net loss" is only a goal, two of the primary regulatory agencies that deal with wetlands under section 404 of the Clean Water Act wrote it into their permitting process. In what quickly became a contested action, the Army Corps of Engineers and the Environmental Protection Agency (EPA) in summer 1990 issued guidelines for achieving "a goal of no overall net loss of values and functions" of wetlands.

This memorandum of agreement focused on mitigating the effects of development in wetlands. Among other things, it set up three steps for the agencies to use in reviewing permits for wetlands projects. The most controversial of these was the requirement for compensatory mitigation to offset environmental damages to wetlands. Mitigation includes restoration of degraded wetlands or creation of man-made wetlands. The memorandum suggested an acre-for-acre replacement as appropriate mitigation, but said more or less may be required depending on the values and functions of the disturbed wetlands.

The memorandum immediately drew fire. Alaska's congressional delegation said the mitigation requirements were too strict and would hamper new development in Alaska. The memorandum finally was amended with a loophole for Alaska that says "avoidance, minimization and compensatory mitigation may not be practicable where there is a high proportion of land which is wetlands."

As evidenced by the memorandum, Alaska no doubt will be treated differently when it comes to wetlands. The Army Corps' regulatory office for the Alaska district is beginning a long-term process of developing a comprehensive mitigation plan for oil and gas developments on the North Slope, a recent directive from Washington. The Corps and EPA also have been approached by oil companies about the possibility of accelerated rehabilitation of abandoned oil facilities to satisfy the required compensation.

Add to this Secretary of Interior Manual Lujan's assurance made in February 1991 to Alaska's Gov. Walter Hickel that Alaska would be treated differently in regard to "no net loss." No one knows what this means in practical terms. Opponents of "no net loss" in Alaska have asked that the state be exempted from any new

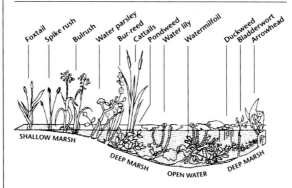

Freshwater Marshes

Freshwater marshes are low-lying areas frequently or permanently flooded with fresh water. They often border streams, rivers or lakes. Freshwater marshland is characterized by "emergent" herbaceous plants such as cattails and bulrushes, which have underwater roots but stems and leaves that rise above the surface. (Adapted from *Alaska Fish & Game*)

policy, but task force members have said an exemption is unlikely.

The federal government clearly faces difficulty in crafting a uniform "no net loss" wetlands policy for the nation. Even a goal of "no net loss" for the nation is beyond the realm of existing regulations: The biggest chunk of wetlands lost each year comes from ditching and draining, actions that are not covered by any law.

WETLANDS INCLUDE SLOUGHS, wet meadows, mud flats, river overflows, swamps and other areas covered by water or that have waterlogged soils for long periods during the growing season. The Army Corps and EPA

define wetlands as "areas inundated or saturated by surface or ground water at a frequency and duration sufficient to support, and that under normal circumstances do support, a prevalence of vegetation typically adapted for life in saturated soil conditions."

Only in the last 20 years or so have wetlands emerged as areas of environmental interest. Scientific studies in a range of disciplines have shown that these wet-soil and shallow-water areas are important in a variety of ways.

Wetlands fill a vital niche in the planet's ecology. They are breeding and rearing grounds for fish and birds, and provide habitat for many threatened and endangered plants and animals. In addition, wetlands are used for recreation, erosion control, floodwater retention, groundwater recharge and act as pollution filters for some aquatic ecosystems.

For many years, however, wetlands were considered unproductive wastelands and their use largely unregulated. As a result, more than half of the

original wetland acreage in the contiguous United States is gone, lost mostly to agriculture, community and industrial development. In the past 200 years, wetlands in the Lower 48 have dropped from some 220 million acres to about 100 million acres; that is more than 60 acres lost an hour, according to a 1990 report to Congress from the U.S. Fish and Wildlife Service. Each year, 350,000 to 500,000 acres of wetlands disappear.

So far, only a fraction of Alaska's 170.2 million acres have been developed, while 22 other states have filled, drained, leveled, plowed and paved more than half of theirs. Of those, 10 states — Arkansas, California, Connecticut, Illinois, Indiana, Iowa, Kentucky, Maryland, Missouri and Ohio — have lost more than 70 percent of their original wetlands. California and Ohio have developed 90 percent or more of theirs.

"NO NET LOSS" IS THE LATEST buzzword in the relatively young pursuit of wetlands protection.

The use of wetlands was not covered by any law until 1975, when environmental groups won a lawsuit against the government that expanded the Army Corps' jurisdiction. Until that time, the Corps had regulated only the nation's navigable waters, as set out in the River and Harbor Act of 1899.

In the 1960s, growing environmental awareness dictated that navigation was not the only criterion on which to base merits of activities along the nation's waterways. A series of court decisions directed the Corps to evaluate permit decisions in broader terms of public interest, weighing public benefits against damages to wetlands, and the agency found itself in the business of protecting the environment.

In 1972, Congress expanded the Corps' jurisdiction to "waters of the United States" with amendments to the Federal Water Pollution Control Act. The Corps interpreted this as the same waters of the 1899 act, but the National Resources Defense Council sued and won a broader definition, bringing thousands of additional square miles under the Corps' jurisdiction.

The first phase of implementing that court decision began in July 1975 and required permits for placement of dredge and fill material in navigable waters plus adjacent wetlands. The Clean Water Act of 1977 replaced the water pollution control act, incorporating regulations for dredge and fill of wetlands in its section 404.

Not until 1979 did the Corps expand its jurisdiction to the tundra permafrost wetlands of the North Slope. To expedite oil development in an environmentally sound manner, the Corps established in 1983 an abbreviated processing procedure to permit certain oil and gas activities, an action that signified Alaska's unique situation.

The concept of "no net loss" emerged in 1987, when the National Wetlands Policy Forum issued recommendations for better wetlands management and protection. The forum was a 20-member group representing state governments, federal agencies, industry, environmental groups and academics, convened by the non-profit Conservation Foundation. President Bush endorsed the "no net loss" goal. He promised the Ducks Unlimited Sixth International Waterfowl Symposium, "It's time to stand the history of wetlands destruction on its head."

Now, the Bush Administration has an opportunity to set the pace of wetlands preservation globally. The United States currently heads the policy-making committee for the Convention on Wetlands of International Importance, which formed in 1971 to stem global loss of wetlands. Its 60 member-nations meet every two years to discuss technologies and emerging guidelines for wise wetlands use. The convention's top priority will be conservation of wetlands in developing countries, according to John Turner, director of the U.S. Fish and Wildlife Service who serves as convention chairman for the next three years.

ALASKA'S 170 MILLION ACRES OF wetlands are diverse, from small ponds, to muskegs with thick trees and moss, to vast stretches of tundra. Tundra wetlands alone total 73 million acres. Another 52 million acres are mostly black spruce bog and coastal marsh. In addition, the state has some of the nation's most extensive complexes of intertidal wetlands.

About half of Alaska's total land is wetlands, including most of the land suitable for development. In addition to covering nearly all the North Slope, wetlands make up most of the Yukon-Kuskokwim river deltas, the Copper River delta, the non-mountainous part of the Seward Peninsula and areas bordering Cook Inlet and the Gulf of Alaska.

However, some of the most ecologically important types of wetlands are scarce. Coastal salt marshes — critical feeding and rearing habitats for fish and staging areas for migratory shorebirds — comprise only 345,000 acres of Alaska's total wetlands. Only about 21 million acres, or 11 percent, of Alaska's wetlands are coastal and deep-water habitat. These types are considered particularly vulnerable to loss because most of Alaska's urban development has occurred in coastal areas where these wetlands are found, according to the National Wildlife Federation. Should Alaska be exempted from "no net loss" policy, "our ability to conserve coastal wetlands habitats vital to maintenance of Alaska's fisheries will be severely compromised," says the National Marine Fisheries Service.

The wetlands in Cook Inlet are limited as well; only 26 percent of the land is wetlands, yet they support a significant part of the continental population of trumpeter swans, the only

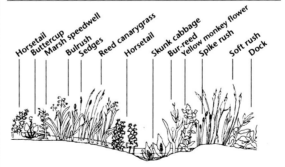

Wet Meadows

Horsetail Buttercup Marsh speedwell Bulrush Sedges Reed canarygrass Horsetail Skunk cabbage Bur-reed Yellow monkey flower Spike rush Soft rush Dock

Wet meadows are free of standing water most of the year. Meadow soil remains waterlogged just below the surface, excluding most upland plants. Wet meadows are often dominated by horsetails, sedges, rushes, skunk cabbage and reed canarygrass as well as wildflowers such as buttercups, monkey flowers and knotweeds. (Adapted from *Alaska Fish & Game*)

known nesting population of tule white-fronted geese and probably are critical to migrant populations of cackling and lesser Canada geese. Likewise, of the nearly 14 million wetlands acres on the arctic coastal plain, less than 2 percent are permanently flooded grass marshes, a principal aquatic habitat for many waterfowl species.

In total, Alaska's wetlands support a staggering number of the nation's migratory waterfowl, including 70,000 swans, a million geese and 12 million ducks. "As wetland destruction continues elsewhere, Alaska becomes more important for duck species in decline," says the National Audubon Society.

In 1980, the importance of many of Alaska's wetlands as wildlife habitats was recognized in a section of the Alaska National Interest Lands Conservation Act. It set up 16 national wildlife refuges of more than 77.3 million acres, focusing land selection on wetlands and the number of waterfowl they supported.

One of those refuges, Izembek, at the tip of the Alaska Peninsula near Cold Bay, is considered a globally important wetland by the international wetlands convention. The entire population of brant, a small goose, occupies Izembek Lagoon in the fall and spring because of this species' dependence on

Many Alaskans depend on the subsistence resources of wetlands for food and clothing. The state's more than 170 million acres of wetlands provide crucial habitat for a number of animal and plant species. *(Staff)*

eelgrass, which is abundant here. Another refuge, Yukon Delta, is the most extensive area of wetlands in the state. Coastal habitats of the delta are critical to perhaps more types of waterfowl than any comparable area in North America, according to U.S. Fish and Wildlife Service biologists. The Yukon and Kuskokwim river deltas together make up the single largest intertidal habitat in the western hemisphere.

Alaska's wetlands provide food and cover for moose, bears, beaver, otter and other furbearers in addition to birds. Moose, for instance, spend hours in knee-deep bogs, feeding on aquatic plants and the various willows that grow on the margins of wetlands.

THE MOST RECENT WETLANDS debate in Alaska is reminiscent of past land-use arguments. Opponents of tighter wetlands controls argue that Alaska has enough land covered by federal and state protected status; any additional wetlands policy is another example of onerous government intrusion.

Stricter wetlands policy, they say, will send development costs skyrocketing and make it difficult if not impossible for cities and villages to expand. "No net loss" will essentially kill the state's economic future in resource development, they say.

The Resource Development Council, one of the most outspoken groups against "no net loss," argues that Alaska is the only state without a wetlands

loss problem, with .05 percent of its wetlands in development, yet it will be disproportionately affected by stricter policies. An October 1990 report to Congress by the U.S. Fish and Wildlife Service estimates Alaska's wetlands loss at less than 1 percent.

Much of what has been written or said about wetlands in Alaska focuses on the impact of oil exploration in the Arctic. The arguments have to do with whether arctic wetlands serve the same functions as other wetlands; whether restoration of wetlands in permafrost is possible; whether mitigation should be required until fish and wildlife populations are demonstrably reduced. Alaska's North Slope accounts for 25 percent of the nation's oil production and at least 85 percent of the state's revenue. Yet oil and gas development, says Steve Taylor, manager of environment and regulatory affairs for BP Exploration in Alaska, has covered only a fraction, less than 20,000 acres, of the slope's 37 million wetland acres. Another Prudhoe Bay-sized oil field, with today's technology, would cover less than 5,000 acres, increasing wetland losses only a fraction of a percent.

"We believe that returning disturbed areas to suitable habitat as rapidly as they become abandoned would achieve the goal of habitat protection without impeding development. In contrast, a no net loss policy would preclude the development of a major new oil field," Taylor says.

Most of Alaska's native corporations also oppose stronger wetlands policy, fearing it could restrict economic development of resources on native lands. Doyon Limited, the regional corporation of interior Alaska, is pursuing oil exploration and mineral development in the Yukon Flats and Kandik River basin as a future economic base for its villages. Doyon, like other native corporations, argues that the Alaska Native Claims Settlement Act set up the 13 native corporations with land expressly to be developed.

Julie Kitka, president of the Alaska Federation of Natives, told the president's task force that "no net loss" could close off growth opportunities for villages facing a population boom, almost all of which are located in

wetlands. Yet the state's largest native organization, she said, is concerned that wetlands be protected because they are important to subsistence. Alaska Natives historically have depended on wetlands. Some 200 rural villages, out of 209 villages counted in the 1980 census, are located in wetlands, along a coast or river. Transportation is easiest here and fish and wildlife plentiful. Wetlands provide an economic base for most villagers.

Generally, local governments and industries such as forestry and mining are protesting the "no net loss" goal.

The Bering Sea Fishermen's Association is one of the few

Guttation water droplets on horsetail (*Equisetum*) look like dew but actually result from excretion of water through the plant's outer covering under conditions of high humidity. *(Jon Nickles)*

industry groups to come out in favor of "no net loss." As the state's second most important industry, fishing offers villages the best way to develop stable economies and increase local employment, the group's president, Jon Zuck, told the task force. "We need strong wetlands protection. . .," he said. "Destroy (wetlands) and we are out of business."

Others supporting the "no net loss" concept say sizable chunks of Alaska's wetlands already have disappeared and such leapfrog development will be devastating if allowed to continue at the current rate. Urban development has claimed 55 percent of Anchorage's wetlands, according to the city's Department of Economic Development and Planning. The loss of some of these wetlands may have contributed to severe flooding in 1989 that caused millions of dollars in property damage. In Juneau, some 30 percent of Mendenhall Valley's freshwater wetlands have been developed, according to U.S. Fish and Wildlife Service estimates.

While only about 200,000 acres of Alaska wetlands have been drained or filled, an uncounted number of additional wetlands may have been

adversely affected through contamination, partial drainage or other changes in the water table, says the conservation group Trustees for Alaska.

Wildlife groups, such as the National Audubon Society and the National Wildlife Federation, argue that Alaska's wetlands provide essential wildlife habitat with the state's permafrost wetlands supporting the world's population of certain species. Additional federal intervention is needed to prevent destruction of these critical resources, they say. Furthermore, an exemption from "no net loss" will send Alaska down the same road to wetlands ruin already traveled by the rest of the nation. "Developers should not be given a blank check to destroy wetlands just because Alaska has more than any other state," says the National Audubon Society.

Some people argue that much remains to be understood about wetland ecosystems and the effect even small habitat losses have on the food chain. Rick Steiner, an associate professor with the University of Alaska Marine Advisory Program in Cordova, invoked the chaos theory when he addressed the president's task force. "Small perturbations can have very large, unanticipated consequences," Steiner cautioned. "The loss of genetic variants can be long lasting."

Environments NOW . . .

POLAR BEARS AND HUMANS

BY SCOTT L. SCHLIEBE

Editor's note: *Scott L. Schliebe is the primary polar bear management biologist for the U.S. Fish and Wildlife Service (USFWS) in Alaska. He has worked extensively with residents of northern coastal villages on marine mammal conservation issues.*

FOR CENTURIES INDIGENOUS people have coexisted with polar bears. These marine mammals have also captured the imagination of the general public. Since the early explorers first encountered it, the great white bear has commanded fascination, respect, admiration and fear. Natural camouflage coloration, stealth and hunting ability, extreme strength, great speed for short distances, intelligence and resourcefulness — adaptations which allow the bears to survive in harsh environments — enhance the polar bear's mystique.

Considering that bears regularly travel along Alaska's

Among the largest of predators, adult male polar bears may weigh 1,500 pounds or more and stand more than 4 feet at the shoulder. *(Steve Amstrup, USFWS)*

northern and western coast, are naturally inquisitive and possess an extraordinary sense of smell, are predators, are subjected to periods of food scarcity and are upper level consumers in the food chain without natural enemies, it seems surprising that more injurious bear/human encounters do not occur. The death of Carl Stalker of Point Lay caused by a polar bear mauling on December 8, 1990, has heightened public awareness of the bears. The public's reaction to the first polar bear-caused human mortality in Alaska in recent time has varied from paranoia to increased respect for the animal. Although chances of polar bear attacks are remote, people must not take for granted that attacks will not occur.

The foremost aspect of the bear's mystique is its ability to kill humans. All North American bear species have the physical ability, provoked or unprovoked, to kill humans. Yet, considering the frequency of human activities in bear habitat, remarkably few people are

injured in bear attacks.

What can be learned from Carl Stalker's death? Dr. Thomas Albert, North Slope Borough Department of Wildlife Management, and Dr. Elizabeth Nelson, North Slope Borough Department of Health and Social Services, conducted a necropsy on the bear and determined: The bear was an adult at least six years old, and as such had been independent for three winters; the animal had essentially no subcutaneous body fat and an almost undetectable amount of interstitial (internal organ) fat; no physical abnormalities were detected which may have affected the ability of the bear to catch prey, such as broken bones, parasites, or damaged, decaying or broken teeth.

Information on polar bear and human encounters was compiled by Susan Fleck and Steve Herrero, Canadian investigators, from 1965 to 1985. Of a recorded 373 aggressive encounters, only 20 resulted in injuries, including six human fatalities. At least 230 bears were killed as a result of these interactions. A typical encounter generally involved a subadult male bear investigating marine mammal food sources at or near hunting camps. The bear's visit usually occurred between midnight and 6 a.m. when people were sleeping. A greater proportion of injuries occurred during the winter at exploratory camps and involved adult male bears. Most neutral encounters involved the presence of food. In more than half of the harmful encounters food was absent, suggesting the victim's presence may have attracted the bear.

The nature of the encounters revealed differences between intentions of male and female animals. Adult male attacks appeared to be predatory and mimicked the way bears strike and kill seals. Bears had to be destroyed to end the confrontation in nearly all cases, and in several cases Fleck and Herrero thought that the bear had actually stalked the victim. Female polar bears, in contrast, did not appear to be attempting to kill the victim but were acting defensively. In most cases, humans who surprised females with cubs seemed to prompt the attack. Seventy-five percent of

Scientists estimate about 3,000 to 5,000 polar bears now range along the northern and western coasts of Alaska. It is difficult to determine the population precisely because polar bears are wide ranging and much polar bear habitat lies within Soviet territory in the western Bering and Chukchi seas. *(Scott Schliebe, USFWS)*

these attacks ended on the bear's initiative. Scientists generally think that polar bears cannot easily be provoked, and will retreat if deterrent devices are used.

A number of biological and behavioral questions arise in trying to determine why a polar bear would attack a human. Since the bear that killed Carl Stalker had no fat reserves, did it reflect undernourishment among the polar bear population in general? Is it unusual to encounter skinny or starving bears in poor physical condition? Are there too many bears, and are they depleting seal populations, locally or rangewide? Are greater numbers of bears frequenting coastal villages in recent times, and if yes, why?

Studies by the Alaska Fish and Wildlife Research Center (USFWS) indicate Alaska's polar bear population is probably at a healthy level, and has increased during the past 20 years although a precise count is unavailable. Conducting population surveys for vast offshore areas is difficult, particularly in the western Chukchi and Bering seas where much polar bear range lies within Soviet territory. The Beaufort Sea stock was estimated at 2,000 bears in 1986. A total Alaska population of 3,000 to 5,000 is projected from average density estimates.

The Chukchi population appears to be stable at relatively high levels. Estimates of population size are not available although a number of factors

indicate population health and vigor: The age composition of harvested and captured bears within the population again includes a representation of old age individuals (absent during the late 1960s and early 1970s); average litter size from animals denning on Wrangel Island, USSR, is high; survival rates of various age classes appear normal; the proportion of adult females with cubs appears representative of a healthy population; overt physical condition looks good (although variable by season and year); Natives of coastal Alaska continue to encounter relatively large numbers of animals in near-shore areas; and there is a slight increase in harvest rates. Of particular interest are St. Lawrence Island communities that encountered polar bears only sporadically prior to the early 1980s; the average annual harvest here in recent years is 30 bears.

Prior to passage of the Marine Mammal Act in 1972, when aerial sport hunting was banned, residents of coastal villages remarked that fewer bears or tracks of bears were seen. Aerial hunters found the majority of polar bears great distances from shore, and scientists thought hunting pressure had altered distributions. Today, polar bears are commonly found along coastal areas, and according to

Biologists use dart guns to sedate polar bears for further study. *(Scott Schliebe, USFWS)*

residents, bears are as abundant as they were prior to aerial hunting.

The animals are associated with the fall advance of ice into near-shore areas. In most years, bears will cross the North Slope in an east to west direction in late October and November. Movement of bears in coastal areas of western Alaska is similarly associated with arrival of arctic pack ice which occurs later here. Food and geographic features also attract bears to near-shore areas. Points and promontories that protrude into polar bear habitat intersect the movements of migrating animals or in some cases result in carcass deposition on beaches. Many times the geographic features that tend to intersect polar bear movements were selected as

village sites many centuries ago because of their proximity to marine mammal food sources.

Interestingly, the Soviet Union is indicating an increased problem of bears appearing in coastal villages. Declining harvests were detected through-out the Soviet Arctic during the 1930s to 1950s. In response to the population declines, the harvesting of bears from ships or at remote polar stations was prohibited in 1938. Since 1956, polar bear hunting has been banned throughout the Soviet Union. Today a limited number of animals, primarily cubs-of-the-year, are taken for zoos and circuses. Strict penalties are provided for unlawful killing of polar bears in the Soviet Union.

The Soviets, through official submission to the International

Union for Conservation of Nature and Natural Resources Polar Bear Specialists Group (PBSG) in 1981, acknowledged a possible increase in the number of bears particularly in eastern Siberia and Chukchi Sea areas. This impression stemmed from data collected from remote weather stations on bear distribution and abundance. The Soviets concluded that bears seem to be losing their fear of humans and that unprovoked attacks on humans are becoming more frequent. The 1985 PBSG submission states:

Our data indicate continuing growth of the animals' population, especially in the northeastern Soviet Union . . . . At the same time, it is becoming . . . obvious that polar bears are losing their fear of humans, as can be seen in their more frequent visits to populated localities (although this may also have resulted from the bears' difficulties in catching seals, . . . due to the severe ice conditions in these seas in recent years). As the situation now stands, it has become necessary to change Soviet Union practice in managing polar bear populations: The authorities have begun restricted selective shooting of individual animals gravitating towards built-up areas and

especially those behaving aggressively to humans. This selective slaughter of polar bears is to be authorized primarily in the Chukchi Autonomous Region, and will not exceed more than a few dozen animals a year.

Physical conditioning is critical to polar bear survival. The ability to hunt is learned, and the first lesson for a cub begins when the mother vacates the den in search of food. Cubs must stay with their mother for more than two years to learn efficient hunting skills. Even then some cubs may be poorly prepared to live on their own when weaned during their third spring. Researchers have recorded remarkable weight gains and losses of several hundred pounds for individual bears captured during different seasons. Conditioning is particularly important for a pregnant female. Some pregnant females have not developed the fat deposits necessary to carry them through a successful six-month, full-term pregnancy. Therefore, some females physiologically discontinue the pregnancy, do not den during that year, and instead breed the following spring and attempt to gain the weight necessary to successfully complete pregnancy.

Even though conditioning is so important, enough individual variability exists that can only be explained by geographic differences in prey (seals), by the movement or population status of prey or by differences in hunting abilities of individual bears. Every year researchers capture skinny bears, although the chances are less that a bear in the fall will be thin as opposed to an animal captured in the spring after enduring an arctic winter. Similarly, a certain number of extremely robust bears are almost always encountered. Polar bears need to eat a seal approximately once every seven days. If seals are unavailable, bears will move in search of pockets of greater seal abundance. Canadian researcher Ian Stirling documented a mass

Polar bears are expert swimmers, having been reported swimming as much as 50 miles from the nearest ice or land. *(Scott Schliebe, USFWS)*

Physical conditioning of the mother plays an important role in birth and survival of cubs. If a pregnant female has not developed enough fat to sustain herself through a six-month, full-term pregnancy, she will end the pregnancy and breed again the following spring. *(Scott Schliebe, USFWS)*

movement of bears out of the eastern Beaufort Sea in the 1974-75 winter because of poor seal production the preceding year and extremely harsh weather conditions during that winter.

It is not unusual to find skinny bears searching for food, generally marine mammal carcasses, onshore or in villages. In the past several years, bears have been observed near or in a number of villages; they have been attracted to the remains of bowhead whale carcasses, have scavenged seals stored frozen outside homes, have been found

in sled dog lots and village dumps, and have dug up fermented meat buried in wooden kegs and meant for human consumption. They have also been shot in the arctic entry ways of village homes.

The situation of problem bears within Alaskan villages does not exist to the degree it does in the Soviet Union since Alaska Natives may legally harvest bears. However, in Alaska hunters in certain villages exercise selectivity in shooting bears. They avoid taking family groups altogether, or target younger animals, sparing the most important family member, the mother. Hunters on the North Slope are implementing a polar bear management plan for the Beaufort Sea region with their Canadian counterparts. Problem

bear situations in Alaska likely would be similar to those of the Soviet Union without hunting, since polar bears have been taken within one-half mile of all major polar bear harvesting villages during the recent past.

How can people reduce conflicts with polar bears in the Arctic? Remain alert for the presence of bears. Develop an understanding of polar bear ecology and behavior. Keep food and garbage secured and unavailable to bears. Anticipate locations bears may use. Develop a mental picture of reactions and escape routes. Learn about deterrent devices to scare bears away if necessary. Maintaining a cautious and respectful regard for polar bears is a healthy attitude which should benefit people and bears by reducing adverse encounters.

ARROWTOOTH FLOUNDER

Arrowtooth flounder, one of the most abundant bottomfish species in the Gulf of Alaska and Bering Sea, may take on new commercial importance as a source of surimi, a white fish paste marketed as simulated crab, shrimp or other seafood. Currently the fish of choice for making surimi is pollock, but stocks of this species are declining, and biologists expect lowered populations to continue. This uncertainty about the supply of pollock creates volatility in the surimi market. To counteract this trend, the Alaska Science and Technology Foundation approved a grant

Scientists are looking to the arrowtooth flounder as a new source of fish for making surimi. *(Courtesy of Diana Wasson)*

Wasson and her colleagues are adding beef plasma powder and egg whites in small quantities to inactivate the enzyme. This procedure enabled them to turn 4,000 pounds of flounder into 400 pounds of surimi. In the future, Wasson plans to blend the flounder with the pollock to produce a stable, abundant supply of surimi.

for the National Marine Fisheries Service (NMFS) to explore the possibility of making surimi from arrowtooth flounder.

Diana Wasson, chemist with NMFS, helped come up with a procedure for processing the flounder. The chief drawback to using this species, which is easily caught in clean tows by a trawler, is enzymatic softening, i.e. the fish's flesh contains a heat-activated enzyme that causes the flesh to break down rather than firm up when cooked. The enzyme prevents the proteins from coagulating, and the flesh turns to liquid.

Arrowtooth flounders, also called turbot or arrowtooth halibut, reach lengths up to 33 inches. They are dark on the upper side of their body, light on the lower side; and as is typical of flatfish, have both eyes on the same side of their head.

LEFT: Employees of All Alaskan Seafoods size grade arrowtooth flounder at their facility on Kodiak Island. *(Courtesy of Diana Wasson)*

RIGHT: A refiner pushes out soft muscle protein in the next to the last step of processing arrowtooth flounder into surimi. Fish skin and connective tissue, undesirable elements, are screened out by the refiner. *(Courtesy of Diana Wasson)*

A PHENOMENON CALLED *AIRGLOW* ILLUMINATES THE night sky in Alaska with a uniform, weak light. Together, airglow and auroral emissions brighten the polar night sky, making the stars appear dim in contrast. Airglow arises from chemical reactions high in the atmosphere, which are triggered by stored energy from the sun. In this *Challenger* photograph looking toward the southern hemisphere, the airglow layer appears as a thin band rimming the Earth's surface. This unusual photo also shows the aurora borealis. (NASA photo by mission commander Bob Overmyer; courtesy of Tom Hallinan, Geophysical Institute, University of Alaska Fairbanks)

ALASKA'S NEWEST HISTORICAL PARK

The state has purchased 13 acres on the shores of Amalga Harbor north of Juneau for its newest unit, Gruening State Historical Park. Ernest Gruening, former territorial governor from 1939 to 1953, non-voting territorial member of the U.S. Senate from 1956 to 1958 and U.S. Senator from 1958 to 1968, began coming to the Amalga Harbor property in the 1940s; the cabin at the site was built in 1949. Here Gruening planned strategy for Alaska's statehood drive, wrote books and entertained well-

(Anchorage Museum)

known leaders in politics and the arts such as John F. Kennedy, Adlai Stevenson and Edna Ferber, author of *Ice Palace.*

Gruening died in 1974, and in 1989 the state purchased the site from his family for $365,000. While the state does not have plans for a full-scale restoration of the cabin at this time, the public uses the property, about 25 miles north of Juneau, for fishing and picnicking. Gruening State Historical Park is reached by a path off the head of the Amalga Harbor boat launch.

OLD JOHNSON TRAIL RENAMED

The trail along Turnagain Arm between Potter and Indian, built during railroad construction in early decades of this century, was officially renamed Turnagain Arm Trail in November 1990. Formerly known as Old Johnson Trail, state park officials who administer the trail in Chugach State Park noted that no evidence can be found that anyone named Johnson was involved in creation of the trail. The name possibly comes from a Mrs. Johnson who, at one time, cooked for railroad crews at the Potter Section House.

Confusion also exists between the Old Johnson Trail and the Johnson Pass Trail in Chugach National Forest on the Kenai Peninsula. Both trails are part of the National Historic Iditarod Trail, but they are not connected.

ANGLER IN SOUTHEAST CATCHES RARE FISH

During the spring sablefish opening in 1990, L. Joyce Davis, a commercial fisher-woman from Sitka, hooked a surprise at 715 fathoms. Davis, who had set some longline on the side of a steep ledge west of Yakobi Island, pulled up a 6 pound fish of a type she had never before seen.

Davis described the fish as smooth, having no scales, having plates instead of teeth, flippers instead of fins, and phosphorous bulbs on its

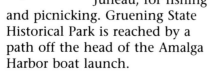

head. The bulbs were glowing when the fish was brought on board but later turned to a gelatinous mass.

The fish was later identified as a giant blobsculpin, *Psychrolutes phrictus.* A rare fish, the first blobsculpin was not caught until the late 1960s. Other specimens have been taken off California, Oregon, British Columbia and in the Bering Sea near Unalaska Island. Davis' specimen may be the first recorded for the Gulf of Alaska. (Taken in part from *Alaska's Wildlife.*)

PRELIMINARY FIGURES ARE IN FOR 1990 COMMERCIAL SALMON CATCH

The 1990 commercial salmon season in Alaska produced a harvest of more than 151 million fish. This total exceeded the preseason forecast of 108.2 million, and makes the 1990 catch second only to the 153.7 million fish taken in 1989.

While some areas posted record or near-record harvests, other areas tallied catches well below projections. On the up side were the sockeye harvest in Bristol Bay and the pink salmon take in Prince William Sound and Southeast. The inshore sockeye salmon run in Bristol Bay produced 47.8 million fish; the pink salmon harvest in Southeast exceeded 31 million fish; and Prince William Sound yielded 44.2 million pinks.

The chum salmon harvest in the Arctic-Yukon-Kuskokwim region, and the pink salmon harvests in lower Cook Inlet, Kodiak and Alaska Peninsula areas did not meet projections.

Preliminary calculations indicate a value to fishermen of the commercial salmon catch at $540 million. 1990's total catch, while smaller than that of 1989, was worth more because of the increase in the valuable sockeye salmon harvest.

METLAKATLA CELEBRATES 100 YEARS

During Tsimshian Days in March, the Tsimshian Indian community of Metlakatla on Annette Island in southeastern Alaska marked its centennial. On March 4, 1891, Congress declared Annette Island a reservation "for the use of the Metlakatla Indians, and those people known as Metlakatlans who have recently emigrated from British Columbia to Alaska. . . ."

The town's history began in the mid-1800s, when Tsimshian Indians from Fort Simpson, British Columbia moved to Old Metlakatla near Prince Rupert. There they lived with their pastor, William Duncan, a Scottish lay preacher sent from London by the Church of England. Twenty years later, when Duncan and other church leaders quarreled, the pastor and some of his followers chose a new home on Port Chester on Annette Island. Duncan remained in the community until his death in 1918, and since then Metlakatla, with an economy of fishing and logging, has grown.

Metlakatla, population 1,386, lines the shores of Port Chester on Annette Island. (Staff)

USSR Connection...

This dairy farm and greenhouse complex is four to five miles from Provideniya at the head of Emma Bay. The Soviets raise Holstein cows, and grow cucumbers in their greenhouse. *(Don Croner)*

CUCUMBERS IN PROVIDENIYA

BY DON CRONER

Editor's note: *Don is a printer in Anchorage who has made several trips to the Soviet Far East.*

FRESH VEGETABLES ARE NOT plentiful in Provideniya, a town of 6,000 people on the Chukotka Peninsula 230 miles across the Bering Strait from Nome. The eastern terminus of the Soviet Union's Northern Sea Route, which stretches 4,000 miles through the Northeast Passage to the Barents Sea in the western Soviet Union, and a distribution hub for surrounding villages, Provideniya appears well supplied with basic foodstuffs like flour, grain, cooking oil, canned fish and hard salami, but perishable items that would have to be flown in from distant agricultural regions are scarce. The town's small grocery store, more the size of an American convenience store than a supermarket, is noticeably lacking in fresh fruits and vegetables.

At a banquet to celebrate the third annual Friendship Flight from Nome to Provideniya the tables were loaded with pastries, plates of justly famous Russian bread, platters of salami and Spamlike meats, caviar, bottles of locally brewed beer, Bulgarian white wine, Georgian champagne, vodka and sweet cakes for dessert. The only fresh vegetables in evidence were plates of sliced cucumbers.

Thirty other Americans and I had flown from Nome to Provideniya on Thursday, June 7, 1990, in five different Bering Air charter flights. My plane left Nome at 2:15 p.m. and landed in Provideniya an hour and 15 minutes later. Because of the time difference between Nome and Provideniya, we arrived at 12:30 Friday afternoon, just in time for lunch. My host, Luba Semenova, 40, led me to her apartment on the third floor of a concrete slab housing complex. She immediately threw out a white linen tablecloth on a hardwood table and served me tea and a bowl of borscht complete with a huge dollop of sour cream. Next came hard

Sheila Davis (left) from Jackson, Wyo.; Olga Lasareva; Irina Anatolovna, who works at the local brewery; and Tanya Lunova, manager of the greenhouse, show off some of the cucumbers grown in Provideniya. *(Don Croner)*

Chukchis, Eskimos and Soviets watch others at a dance outdoors at Provideniya, a community of 6,000 on the Chukotka Peninsula in the extreme northeastern corner of the Soviet Far East. Note the tattooing on the face of the Eskimo woman at left. *(Don Croner)*

salami, salmon roe, bread and butter, *pelmeni* — a cross between ravioli and dumplings — and a plate of sliced cucumbers.

After lunch Luba took me for a tour of the local clothing, grocery and hardware stores, the huge indoor swimming pool filled with heated sea water, and the book shop, where the salesperson insisted that I accept a complimentary poster of Mikhail Gorbachev. We finally caught up with the other Americans at the Sports Complex, a large building equipped with basketball and soccer courts, weight-lifting rooms and meeting halls, for a short welcoming ceremony. Then we all trouped to the banquet hall where, after some entertainments by schoolchildren and the prerequisite speeches and toasts, we finally got down to what the Soviets clearly considered the evening's real purpose, dancing to thunderous Russian disco music.

The next morning, Luba, her son Sergei, 18, (her husband Yuri was working out of town), and I sat in her cozy little kitchen and drank tea prepared Russian-style (a large amount of tea leaves steeped to make a strong, potent decoction that is then diluted to taste with hot water, a method still practiced by many Russian-Aleuts on Kodiak Island and the Kenai Peninsula). Luba quickly whipped up scrambled eggs on her double burner hot plate, and we ate them with bread, salmon roe, salami — and a plate of sliced cucumbers.

Luba then walked me to a bus stop where, with six other Americans, I clambered on board one of the small, 15-seat Russian buses that constitute the main means of local travel. I saw only two automobiles in Provideniya, both of which reportedly are owned by the Communist Party. We left town and bounced our way three or four miles along Emma Bay, the fiord on which Provideniya is located. On the grassy flats at the head of the bay grazed a dozen black-and-white Holsteins, the same kind of dairy cows raised in Alaska. We pulled up to a complex of buildings, including a long, low cow shed and a huge greenhouse. "This must be where they raise all the cucumbers," I said jokingly to John and Sheila Davis of Jackson, Wyo., who had accompanied the mostly Alaskan party. "Did your host serve you cucumbers too?" asked Sheila. "We had them for breakfast."

As we piled off the bus, I indicated to some of the Russians by sign language that I would like to tour the low barn, which was apparently where the cows were milked. This elicited a chorus of *nyets* along with much comical pointing at my shoes. They obviously did not think my Reeboks were up to the job. I was well aware of the hazards of walking through barns, but not well enough equipped linguistically to argue the point.

Instead we entered the greenhouse. I was momentarily taken back. The building, perhaps 200 feet long and 60 feet wide, was taken up almost entirely by cucumber plants. This was indeed the source of Provideniya's ubiquitous cucumbers. We were met by the manager of the greenhouse, Tanya Lunova, and her son Slava. Tanya produced a huge glass jar of fresh, warm, whole milk and poured everyone a big glass.

After touring the greenhouse, we went outside for a mid-morning snack — huge chunks of reindeer meat on long wooden skewers cooked over a barbecue pit. Walking around the grounds I discovered another greenhouse, just a small room actually, located on the second floor of an old building and accessible only by a precarious ladder. This greenhouse contained lettuce, tomato plants, dill and some purely decorative flowers.

My first host, Luba, had to leave town for personal reasons, and I was entrusted to the care of Tanya Lunova. The next morning I ate breakfast at her apartment. We had thick, deliciously oily pancakes served with sour cream instead of syrup, bread and butter, fried salami, a salad of lettuce, tomatoes and fresh dill — and a plate of cucumbers. After I finished the cucumbers Tanya asked, "Don, you like cucumbers, yes?" I answered quite truthfully that I love cucumbers. "Is good," replied Tanya. "We have many cucumbers."

Northern Ink . . .

THE EYE OF THE NEEDLE, retold and illustrated by Teri Sloat, based on a Yup'ik tale as told by Betty Huffman, Dutton Children's Books, New York, 30 pages, hardcover with jacket, $13.95.

Editor's note: *Bob Henning, president of The Alaska Geographic Society, notes that Betty Huffman is the daughter of Mr. and Mrs. Joe Jean of Mumtrak down in Goodnews Bay. Betty's mother was Eskimo and passed on many a story to Betty, who for some years was head of a bilingual institute at Bethel.*

When a child asks to read a book over and over, it is a

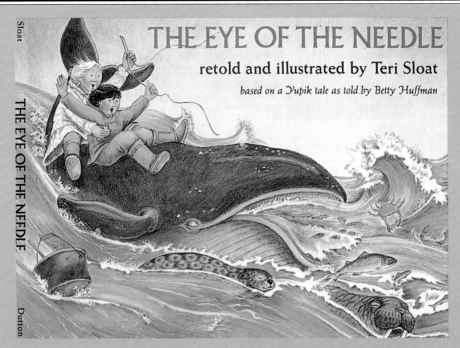

sign that something in the book works. That is how it is with Teri Sloat's newest children's story/picture book, *Eye of the Needle*. It is colorful

and fun to read aloud, with words like "gulluummp" sprinkled throughout.

The story is about Amik, an Eskimo boy, who lives with his grandmother in a sod hut on the edge of the Bering Sea. It is the beginning of spring, and she sends him out to find food. He brings home his catch to her, but in a most unusual way. He swallows a great whale,

"gulluummp." Then the magic starts, and Sloat's drawings take on their own momentum. Grandmother's ivory needle comes into play, and Amik's catch floods the pages in a swish, swoosh, splash, whooshing rush of hooligan, salmon, seal and walrus, enough to feed an entire village.

Sloat, who wrote and illustrated *From Letter to Letter*, a

New York Times Best Picture Book for 1989, based *Eye of the Needle* on a traditional Yup'ik tale told by Betty Huffman. Sloat, now of Sebastopol, Calif., met Huffman, a Yup'ik, during the 11 years she taught in southwestern Alaska.

This book has several things going for it. The illustrations are realistic and fantastical, with historically appropriate details. The pictures, particularly those of the animals, consistently elicit comments and questions from my 2-year-old. Older children may have fun latching onto key phrases that punctuate Amik's eating spree. A quality touch is the

illustrated cover; jackets don't usually stand up to a toddler's handling so an illustrated cover is welcome.

—*L.J.Campbell*

ALASKAN AVIATION HISTORY, 1897 TO 1930, by retired Alaska pilot Robert W. Stevens, Polynyas Press, Des Moines, Washington, 2 volumes, 1,095 pages, more than 980 photographs — many never before published. Hard cover, $150 plus $5 shipping.

We have written this much over the years ourselves, we can presume, but to see it all in one chunk is a little overwhelming. We think it was Horace Greeley (If he, Mark Twain and Will Rogers had really said all the things for which they are quoted, they must have been motormouths!) who once reviewed a book with the simple declaration that "It weighed 4 1/2 pounds." Bob Stevens' two books have a total weight of 9 1/2 pounds, and I am tempted to leave it at that, but I guess Guinness has bigger pounders that than anyway.

But this monumental effort deserves more than *Gee Whizzes*. It represents 15 years of painstaking research. Many of us in the Alaska word business have tackled this one in bits and pieces, some of us in pretty fair books, but Bob Stevens is the first one to come along who really went that extra mile to record the stories and place in record the names of pilots mostly forgotten, of planes and

Robert W. (Bob) Stevens poses at Anchorage on May 19, 1978, prior to his last flight before retirement. *(Courtesy of R.W. Stevens)*

incidents in northern aviation history that to all intents were essentially unknown.

Few realize that much of the history of world aviation itself was written in Alaska, and many of the almost-forgotten aviators of the North accomplished unbelievable feats of pioneer flying and even early airplane design and construction.

No person who enjoys a book on the history of flying should miss this. We are not going to attempt to name names and events. The review would quickly achieve its own record weight.

But take it from me. . . I knew many of these people and they would appreciate the job Stevens has done. Noel Wien would have said, "Good book!"

When you read it, as you must if you are at all turned on to either Alaskan history or Alaskan aviation, its' 9 1/2 pounds become a ton in contribution to either of those categories.

The inevitable question, of course, is when can we expect another few pounds of book to take us up through the middle years of Alaskan aviation that came to a close for the most part with the arrival of the jet?

Bob, whatever awards or rewards you may earn from *Alaskan Aviation History*, we add our own. . . "The best early Alaskan aviation history ever written."

Order from Polynyas Press, P.O. Box 98904, Des Moines WA 98198.

—*Bob Henning, President*

Published by
The Alaska Geographic Society

Robert A. Henning,
PRESIDENT
Penny Rennick,
EDITOR
Kathy Doogan,
ASSOCIATE EDITOR/PRODUCTION COORDINATOR
Kaci Cronkhite,
DIRECTOR OF SALES AND PROMOTION
Lori Granucci,
MEMBERSHIP/CIRCULATION ASSISTANT
L.J. Campbell,
STAFF WRITER

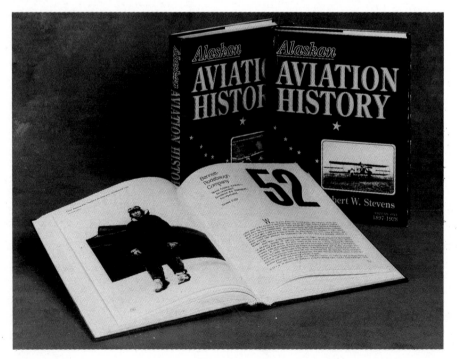